Fire In The Hills

A Collective Remembrance

Edited by

Patricia Adler

Marion Abbott Bundy
Linda Morris Fletcher
Nancy A. Pietrafesa
Terry Shames
Jane Staw

For information, write to: 2904 Avalon Avenue, Berkeley, CA 94705

ISBN: 0-9634167-0-7
Library of Congress Card Catalog Number: 92-096892

The editor and authors are grateful to the following for granting reprint permission:
M Magazine, March, 1992, "Deal With It," Jeremy Larner
The Pacific News Service, October 23, 1991, "Outrunning the Firestorm," Deirdre English.
Phoenix Journal, January-May, 1992, "Fire Journal", Tobie Helene Shapiro;
"The weekend After The Fire," Judith Stronach; "First Cookies", Elsabeth Wickett; "Zones," Constance R. Rowell.

Cover photograph: Fire in the Hills, Richard Blair

Creative Direction: Pennfield Jensen
Design, Production & Color Separations: Venkatesh
Printed through Penn & Ink by Colorcraft, Ltd., Hong Kong

Proceeds from sales of this book will be used to benefit the Alta Bates Burn Center.

CONTENTS

III

IV

PHOTOGRAPHS

Cover: Fire In The Hills, *Richard Blair*—Photo taken at about 2:30 p.m. from a rooftop on Procter Street
Back cover: After the Fire, *Constance R. Rowell*

Color Plates I:
Daughter on Roof, *William Fletcher*
Making a Stand, *Sean Bonetti*—Photo taken at 4:30 p.m.
Alta and Agnes Streets, *Sean Bonetti*—Photo taken at 4:45 p.m.
275 Alvarado, *Steven Frus*
Untitled, *Steven Frus*
Helicopter, *Trent Nelson*
Fire in the Hills, *June Felter*—Artist June Felter photographed views from her hilltop house instead of
        saving her life's work of watercolors, drawings, oils and acrylics dating from 1960.
Proofsheet, *Laurie Dornbrand*

PREFACE

I n those first hushed days after the containment of the October fire, the most ordinary routines seemed remarkable. Morning newspapers landed on sidewalks at daybreak, at addresses where houses no longer stood. Garbage trucks lumbered down streets early for their weekly collection, their routes shortened. And faithful UPS wagons circled, desperate to make promised deliveries.

The simplest chore felt like a rare privilege. That some of us still had beds to make, desk work to finish, refrigerators to stock, roofs to repair was miraculous. A trip to the market or to the local gas station became a community event. Whose house was left? Whose was not?

A stop at the post office was a shock. Long lines of the newly homeless waited to apply for postal boxes or to obtain change of address cards. Strangers and acquaintances passed the time telling their stories to each other with brave humor and unabashed sorrow. Laughter over borrowed clothing or bizarre living arrangements faded into low whispers of concern for neighbors still missing. People rode a roller coaster of disbelief, clustering round, trading tales and feeling more like a community than they ever had before.

Our stories, whether of total loss or freakish survival, were spellbinding. One story unleashed another and another. The first giddy sagas of heroism and escape gave way to the trials of the aftermath. One would-be renter lost seven rentals to higher bidders in the time it took to conclude a telephoned agreement and to arrive with a check of deposit in hand. She finally succeeded on her eighth attempt.

It is now July, and still the stories keep coming. Stories are one thing the fire couldn't take from us, and those it gave are some of the most vivid and clarifying of all. The stories in this collection are not simply about a fire but are also about courage, humanity and resilience put to extreme tests. As our tales unfold, we see how we have been changed and, in unexpected ways, been renewed.

Stories endure. They link us to our past and future. When a disaster strips us of everything else, our stories become signposts. They help us find meaning, rediscover who we are, and move on. There are many stories yet to tell.

Making this book has been a communal act. Despite the sorrow of its contents, there have been moments of great joy in collecting these remembrances and in receiving the help and guidance of so many open-hearted backers. My deepest thanks go to all of you who lit the way:

to the writers, artists, rescuers, and survivors who responded to our call;
to our advisors and supporters: Ronald Adler, Dale Block, Stephen Bundy, John Danner, William Fletcher, Ray Gatchalian, Bonnie Glaser, Pennfield Jensen, Dennis Kuby, Margaretta Mitchell, Marian O'Brien, Dick Schuttge, Deborah Simpson, David Shames, Dorothy Stein, Venkatesh;
to our production assistant, Katherine Falk;
to those we interviewed: Robert Bruce, Steve Carvahlo, Lennie Fisher, Bob Gattis, Julie Lehman, Don Pearman, Nicole Schapiro, Doll Stanley;
to our animal companions, friends and family members who know our stories and keep listening

Patricia Adler

OVERVIEW

Late Sunday morning, October 20, 1991, a fire of unprecedented force blew out of control in the Oakland-Berkeley hills. By nightfall, 3,354 houses and 456 apartments had been destroyed, 5,000 people were without homes, 150 people injured, and twenty-five people dead.

A confluence of circumstances and events conspired to produce this disaster. California was in its fifth year of drought. A severe freeze the previous winter had killed or weakened vegetation. The terrain at ground zero was very steep, making it difficult for firefighters to reach the fire, and producing strong updrafts to feed the flames. Many of the houses had shake roofs, and some of the streets close to the fire's origin were so steep, narrow and winding that fleeing cars created bottlenecks.

The fire actually began the day before. A brush fire erupted behind Buckingham Boulevard in Oakland, near the Caldecott Tunnel. At the foot of the hill, the University of Washington Huskies were beating the California Bears in the Cal football stadium. The day was warm and dry, but not unusually windy. Firefighters battled the brush fire for about two hours and "mopped up" until dark, when they left the area, their hoses still in place. The cause of the brush fire has not been established.

On Sunday morning, Oakland firefighters returned to gather their hoses. They noticed hot spots—embers still alive and occasionally flaring. The morning was hot, and at about 10:00 the wind started picking up, a "wrong-way wind," blowing like a blast furnace from the valley in the east. Some people call this the "Santa Ana," for it mimics the famous desert wind of Southern California.

By 10:30 the temperature was 90 degrees, the humidity 16 per cent. The winds were now gusting up to 50 miles per hour, acting as a bellows on the smoldering embers. Firefighters put out small blazes. Suddenly the whole hill burst into flame, rushing upward in a wall of fire that created its own internal winds.

Chaos followed. Dispatchers were overwhelmed, communications erratic, equipment overtaxed. In the first hour of the fire, one house went up in flames every 4.5 seconds. Some people were warned; some were not. The fire reached 2000 degrees. By noon, 700 houses were gone. Churning black smoke was seen for miles.

In addition to many volunteers, almost 2,000 professional firefighters fought the blaze. Helicopters and airplanes were called in. Reservoirs ran dry. The fire died down only when the wind did.

By midnight, gas-jet flames dotted the blackened hills where the houses had been. Over 1,600 acres had burned. Hundreds of pets were lost. Property damage was estimated at $1.5 billion.

The fire has been called an "interface" fire, one that burns where wild lands abut populated communities, calling for both forest and urban firefighting techniques. It may also be the "fire of the future" for California.

IN MEMORY OF

Eunice Barkell, 79, of Charing Cross Road,
Oakland. Found in house on 6677
Charing Cross Road

Gail Baxter, 61, of Buckingham Road, Oakland.
Found in road on 6800 block of Charing
Cross Road

May Elizabeth Blos, 85, of Live Oak Road.
Found in house on 100 block of
Live Oak Road

Mary Lucile Brantly, 78, of Binnacle Hill Road,
Oakland. Found in her home, which is
in middle of block

Robert Emery Cox, 64, of Kenilworth Road,
Berkeley. Found in backyard on 100
block of Chancellor Place near corner of
Drury Road

Terry DuPont, 58, of Kenilworth Road,
Berkeley. Found with Robert Cox

Carolyn Grant, 75, of Marlin Cove Road,
Oakland. Found in remains of house
on 100 block of Marlin Cove Road

John Alexander Grant, 77 of Marlin Cove
Road, Oakland. Found with his wife,
Carolyn, in their house

John Grubensky, 32, of Linden Avenue, Fairfield. Found on 6800 block of Charing Cross Road about 15 feet uphill from Gail Baxter

Segall Livnah, 18, of San Diego. Found in street about mid-block on Windward Hill

Phillip Loggins, 51, of Charing Cross Road, Oakland. Found on Charing Cross Road about 60 feet uphill from John Grubensky

Lucy Chi-Win Mantz, 46, of Schooner Hill Road, Oakland. Found in remains of house on 100 block of Schooner Hill Road, near corner of Charing Cross Road

Gregor McGinnis, 46, of Bristol Drive, Oakland. Found in backyard of house on 6800 block of Bristol Drive

Lewis D. McNeary Jr., 44, of Charing Cross Road, Oakland. Found on Charing Cross Road, just above Tunnel Road

Patrick Emmett O'Neill, 40, of Norfolk Road, Oakland. Found on Bristol Drive at Buckingham Boulevard

Leigh Ortenburger, 62, of Los Palos Avenue, Palo Alto. Found in driveway of house on Charing Cross Road near Sherwick Drive

Martha Gabriela Reed, 18, of Orinda. Found on Charing Cross Road, 40 feet from intersection of Schooner Hill

James Riley, 49, of Silver Lake Drive, Martinez. Found in middle of the 6900 block of Norfolk Road

Kimberly Robson, 37, of Buckingham Boulevard. Found in driveway next to James Riley

Francis Gray Scott, 37, of Alvarado Road, Oakland. Found in house on 1000 block of Alvarado

Virginia P. Smith, 61, of Charing Cross Road, Oakland. Found in driveway on Charing Cross Road, above Tunnel Road

Anne Tagore, 54, of Norfolk Road, Oakland. Found in house on the 6900 block of Norfolk Road

Aina Turjanis, 64, of Charing Cross Road, Oakland. Found in middle of 6800 block of Charing Cross Road, just east of Sherwick

Cheryl Turjanis, 25, of Charing Cross Road, Oakland. Found 10 feet downhill from Gail Baxter on Charing Cross Road

Paul Tyrell, 61, of Bristol Drive, Oakland. Found in his pickup truck behind John Grubensky's patrol car on Charing Cross Road

I

PREMONITION
for Nancy Pollock

You knew,
even before the hills
were thoroughly dry.
The high grass
knew it too . . . something
about the way the weeds stood
so utterly still, as if listening,
as if hearing already
the rabbits screaming
when the fire finally did
swing down from the ridge
driving them into tighter and tighter
circles of frenzy.

Somehow
you knew, even before
it exploded late summer's languor.
It was as if the tall weeds
heard already
the cries of the maddened blue jays
when, later, they darted blindly
out of the black smoke
and plummeted to the earth,
their wings burning.

—*John Pollock*

On the morning of the fire, Laurie Dornbrand took her son, Aaron, to Children's Fairyland. Just before leaving her house, she impulsively pulled a camera from a drawer. She finished the roll on October 22, the first time she and her husband were allowed back to the site on Grand View Drive where their house once stood.

II

Outrunning the Firestorm

Deirdre English

It happened so fast. Five minutes later and we would all be dead.

There had been no warning, no sound of fire sirens, no squad car commanding an evacuation. At 11:00, I was making breakfast for house guests when we smelled smoke and saw the sky darken. At 11:30, we were running for our lives.

Towers of flame were bearing down on us from three directions, and we had nothing left to do but abandon our cars and outrace the fire as it closed in behind us.

Our neighborhood was one of the first areas hit by flames. There was nothing but dry hillside between the spot where the fire began and my street, and in the minutes it took to sweep down that hillside, the flames had already become a firestorm.

At 11:00, my visitors—my best friend's son, Ben Ehrenreich, and his two best friends, Adam and Jeff—and I went outside to see where the smoke was coming from. We saw black plumes rising from over the hill. It still seemed far away.

I called the fire department. "Yes, we know about it, firefighters are on the scene," said the voice on the other end of 911 in a reassuringly bored tone. I called a neighbor to be sure she was aware, and got her answering machine. "Guess you're not home. Well, I'll come up and hose down your roof if it gets any smokier around here," I said lightly. I hung up and went back to thinking about breakfast when the phone rang. It was my husband, Don, who had left the house about a half hour earlier and was near downtown Berkeley. He had seen the smoke and wanted to make sure I had. "Are you kidding? Of course! I've already closed all the windows. I'm getting a little concerned."

Don advised me to prepare to evacuate. "Get the car and put the valuables in it just in case. I'm on my way home," he said.

Even though it seemed premature, Ben brought the car to the front door, and I asked Adam and Jeff to carry out my most cherished possession, a framed work of art of mostly sentimental value to me, inherited from my father. By now ash was falling through the smoke, and the air was rapidly getting darker. A sense of crisis set in.

The boys threw their backpacks into the car, and I frantically ran into my office and grabbed a file of documents and an armload of photographs off the shelf. I remember pausing for one moment of paralysis, wondering what I was forgetting to take. Then we all jumped in the car and took off.

We got only as far as the corner. I swung to the left, and saw nothing but dense black smoke and shooting embers ahead. I backed up and tried going to the right. The hillside on both sides of the road was aflame, and burning branches were falling onto the street. It was a life-or-death decision: Gun the car and drive past the flames, into what? Or break and back away, admitting that we were trapped. I froze for a moment and heard Ben yelling, "No, no, Deirdre, don't go." In the same moment I knew it was too dangerous to drive. We had to run.

Two other cars containing neighbors also stopped and emptied. Now there were nine of us bolting towards the only way left to escape. It was downhill, down the upwind side of the hill. There was no path, just a steep open field of dry brush. We ran under a tree limb and over a fallen-down fence, staying together as a group, helping each other and yelling to each other as we fled. To my horror, the hillside had already erupted in flame not more than 15 feet away. Now that I had witnessed the speed and fury of the firestorm, I knew that in moments we could be engulfed.

We emerged on Grandview Avenue where people were evacuating but still totally unaware of how fast the fire was converging on them. I tried to flag down two cars which passed us, crammed with people, uncomprehending how dire the threat was to anyone on foot. A third car stopped, though it too was full. I shouted to the three boys to leap onto the trunk. They held on, and it sped away.

All of the others in our group were picked up one or two at a time by residents fleeing in cars. A man came out of his house to escape in one car and threw the keys to his second car to my neighbor Tony and me. We tooled out of the danger area in his gold Mercedes Benz. I rolled down the window and called out to all the relaxed people I saw— some of them simply standing and looking skyward at the smoke: "Houses are going up; time to go! Evacuate! Evacuate!" I felt like Cassandra. It was as though I were driving backwards in time to a moment of calm that I had lived a half an hour earlier.

At the bottom of the hill, on Claremont Avenue, I found my husband, and parted with the Mercedes and my neighbor Tony, an artist who had left behind all his oil paintings.

My neighborhood happened to have been home

to a number of writers and artists whose lifework has been destroyed. A handbuilt organ, a collection of original photographic prints, a manuscript of a book in progress, the master tapes of a radio play—all this and more was lost within the space of a block.

It doesn't matter. All that matters is life, surviving. I have flames under my eyelids when I try to sleep at night. But by day I am filled with the joy of gratitude, the thankfulness that life is given to us, and an increasing sense of unreality that there ever was a moment when I felt like a moth in a fireplace.

THEN IT RACED TOWARDS US

then it raced towards us
down from the hills
we were naked sunning ourselves
we had just washed the cars
illegally in the drought
and the hose still lazed
in guilty pools
but as the sparks
flew into our yard
we were ready
the white
of the Mercedes
neon in the black wind
the birds
iridescent as fish
flying into the red sun

—*Ellen Cooney*

FIRE ENTERS THERAPY

unappointed, crashes thru fine oak
front door, not middle-class
decent enough to refrain
from barging into Dr.'s
office at the exact
moment he's preparing
to trisect a failed novelist's
dream of nearly
successful orgasm. Sits
itself on the wine
stained leather sofa
reducing the neurotic
novelist to pure
ash
& unfolds, blue
petal after blue,
& offers a life
time of moderate consumption—but
Oakland enters
its blue heart
during a night
mare & a hitherto
unacknowledged greed rises

—a secret flame within its flaming
heart—&
whole hills
get swallowed
& as the Dr. swallows & prepares
to steady himself
realizes his tool/kit
of platitudes FINALLY won't
work &
so joins his
diplomas
up in smoke.

—*Norman Weinstein*

Mother Love
Ann Leyhe

A decade ago my mother was diagnosed with advanced breast cancer. My cousin, who is a physician, was joking with her once and said she was "one in a million." When she demurred, he said, "OK, one in 500,000." I think he is probably right.

Nine years ago, it seemed likely that she would never know her grandchildren. Now she has seven of them living here in the Bay Area. My siblings and I never expected to live near each other, but we've all moved here to be close to Mom.

Living near family has its inherent stresses. It helps that my sister Claire and I hike the Strawberry Canyon fire trail every Sunday. We get a chance to talk (mostly family matters, of course) and to get some exercise.

The morning of the fire we made about half our usual distance before the ferocious winds pushed us back down the canyon.

"This is fire weather," Claire said.

While we were driving home, everything suddenly turned mustard yellow, as if we'd donned ski goggles. People drove erratically. A twenty-foot wall of flame burst over the ridge at the top of Tunnel Road.

Claire shouted expletives—not at all like her. What should we do about Mom?

Just three days before, Mom had moved from north Berkeley up to Roble Road. I turned up Tunnel past the Claremont, and right onto Roble.

Mom's new place is one of those irritating '50s houses that gives away nothing from the street. With the drapes drawn and the garage door down, it's impossible to tell if anyone is home. Claire dashed out to pound on the front door while I idled in the middle of the street, afraid that if I pulled over we would be unable to get back into the line of fleeing cars.

Embers swirled through the air. Pine needles caught fire down the street even as neighbors stood talking. Get going! I wanted to scream. Claire jumped back in the car, breathless. She'd tried everything. Mom didn't seem to be there. But her hearing isn't the best. Now what? No police or firefighters anywhere. All we could do was head home.

Part of a terrifyingly slow caravan, we headed down Roble toward Chabot, through a shower of sparks, praying the gas tank wouldn't blow up. When it was safe, Claire bailed out on foot, figuring she could reach her home in Rockridge faster.

By the time I made it back to Webster Street, my husband had raised Mom on the phone. She hadn't heard Claire banging; she told him she would grab a few valuables and leave.

I panicked, thinking of her likely escape route—the way Claire and I had come would be engulfed in flames by now. Tunnel Road would be closed too. I doubted Mom was strong enough to walk out alone. I saw myself in that scene from *Gone With The Wind*, searching for Mom in the ruins.

Instead, she pulls up in front of my house, in her white Honda, herself a vision in white: white hair, white clothes. She waves that kind of wave beauty queens use on floats in small town parades.

Mom had simply glided down a deserted Tunnel Road, the wrong way.

Deal With It

Jeremy Larner

n the Saturday before the fateful Sunday, I jumped the one-hour commuter flight south for what I imagined would be a playful L.A. weekend with a woman friend. I expected to pick up our romance, but soon learned I'd fumbled. There I was, at an electric folk concert next to a lady so less than glad to see me she would neither speak, touch nor make the most trivial eye contact.

For the space of several spunky, unfelt ballads, I occupied myself with what turned out to be the first of many reflections on the little surprises of life. In search of the quickest segue to happier times, I opened with a piece of standard dialogue. "What's the matter, honey?" I said, which I already knew to be, in such cases, the likely equivalent of "kick me." My intimate friend so passionate of late by fax and cellular phone turned a half-profile of controlled fury. "You can't possibly know," she muttered stage left, "the garbage I had to shovel this week, arranging fucking fifteen-minute interviews in New York. Some cretins never stay bought. This is nothing personal. Deal with it."

A veteran of scripts, I recognized the opening of an S-M movie with a 3:00 a.m. pseudo-reconciliation quasi-climax scene, followed by existential verite in the morning. Rather than spend my weekend as a casting error, I dealt with it by renegotiating the maze of LA, trading my rented vehicle for a plane to SFO. In the course of two hours I went from an amphitheater behind barbed wire on a movie lot packed in among pop music cretins, to my strategically-placed canyon home in the East Bay hills, a country setting of winding roads without sidewalks, stars shining clearly in the night and fresh air smelling of pine trees. What magic to make three or four quick secret turns and go from the grubbiness of Telegraph Avenue to a tower of tranquillity only five minutes from the best yuppie restaurants of Berkeley, fifteen minutes from the Oakland airport and twenty from San Francisco. As my genie garage door slid securely down behind me, I entered my dining room from the garage, punched off my alarm, and gazed briefly from my upper deck at the lights of the Bay. I then descended the stairs to a

night of sound sleep in my bedroom cantilevered over an emerging underground creek, whose waters gurgled from the hillside with just enough music to cushion the distant freeway hum.

Two years ago my previous house in the north Berkeley hills was half-destroyed by a fire resulting from tacked-up wiring which shook loose in the '89 earthquake. Every time I walked my dog on the more densely wooded south ridge where I'd moved, I thought how the dead clumps of brush and eucalyptus would flare in a fire. As we moved into yet another drought year, hill and canyon dwellers all over the state made nervous fire jokes while waiting for the rainy season to come.

So there I was, in my sanctuary on Grand View Drive, high above the humdrum but close enough to sample it, planning a day of rest among the books, CDs and personal items I had collected about me, and which ordinarily could be counted upon to remind me who I was and what I might be up to, regardless of how the inconvenient others of the world might treat me. I had never lived in a place so cozy, so easy, so amenable to the illusion I had things in control.

On Sunday, October 20, I awake in my very own bed, but in an uneasy mood for a man coming off an act of liberation. As I haul myself up to the street in my pajamas for the Sunday paper, the day feels subtly unsettling, not unlike quake day in October of '89. A few days before, we had experienced the oddity of 96 degree heat, almost as rare as a lightning storm out here. This morning the air is staled by a hot west wind blowing from the dusty valleys and tugging at my cuffs in swirls, a kind of scirocco that makes the teeth clamp. I fall back in bed and drift to sleep again reading the funnies.

I would not have been surprised to know that the day before a brush fire had flickered a mile above me, just beneath Grizzly Peak Boulevard, the curving, often fogged-in summit road from which Tilden Park runs down the eastward side of the hills towards the elite bedroom suburbs of Orinda and Lafayette. At dusk and again at dawn, a truck had been stationed on a fire trail, manned by several firefighters, including a captain. At some time during my morning sleep, the same fire flared right under their noses, mushrooming out of control in a gust of wind.

Around noon I get up again to let my dog out. I stand outside, thinking I've rarely seen her act so spooked and whiny. I am still in pajamas, perhaps an odd sight to the cars that now and then wind up the hill. I smell a whiff of smoke, and inwardly argue against what I have come to think of as paranoid fire thoughts. Someone must be burning leaves on a fall Sunday. I think I'll take a long bath, then maybe a walk, find out in passing where the smell comes from, not yet knowing it will stay with me awake or sleeping for the next two weeks.

I step out on the back deck for the view along the creek bed. No smoke. At the downhill end of the creek, to the southwest, I can make out Lake Merritt and downtown Oakland through a break in the landscape. Beyond that the Bay, and across it I can see the stretch of landfill off San Mateo, where the 49ers are about to kick off at Candlestick. The sun stands high in the southeast, shining dully through haze in a cloudless sky. Listlessly, I sway towards my hammock and almost drop into it, hanging for a moment in a state of dislocation.

With startling suddenness—as fast as it takes to tell—coarse brown smoke spews up into the sky and covers the sun. Instantly, my thoughts stop.

I enter my bedroom, on my way, it seems, to squeeze in a quick, bracing shower. But no—I pull on swimming shorts and a rayon shirt, trot back up to the street.

I find my organized and conscientious neighbor George, who works for IBM, standing on his rhododendrons spraying water onto the roof of his house. "What's happening?" I say, in my pleasant neighborly style, designed to bring the relaxed answer, nothing much. But usually helpful George keeps his eyes on his roof. "Major fire," he says. "It won't get over here," I tell him.

George makes no reply. Grudgingly, I pull out my own hose. My nozzle isn't as good as George's, and my roofline higher; my spray can't reach much of it. In the distance I see figures standing on rooftops. I drag out a stepladder that doesn't quite reach, haul myself onto the shingles with my free hand and slip into a slow dreadful slide. The hose lashes and spits as I catch the gutter with my free hand, feeling it almost give. Dangling ridiculously, I probe with one foot for the top of the ladder.

For the first time I realize a fire is really on the way. How humiliating, to be carted off with a broken leg while my house goes up in smoke! In an act of supreme will, my toes find the ladder.

Darkness spreads as if a storm is about to break. I tug the hose down the side stairs, spraying. So much house and such an ineffectual splashing of water. From the corner of my eye I glimpse a black and red silhouette leaping between the hills to the east. I give up watering. The stuff inside is more important than the house, I know that. But which stuff? I look around at improbable objects, my big cherrywood desk, my orthopedic chair, my six new hi-fi speakers anchored on various ceiling beams. I pick up my fax machine and it falls to the floor in separate pieces.

It's easy enough to dash out with my toilet articles and half-unpacked overnight case. Later on I find the case contains two books and one towel.

Getting serious, I back my car out, turn it around and leave it in the driveway facing the street. This takes a while, because I must interrupt the steady stream of traffic, all of it uphill, in the direction of the fire. The cars and motorcycles, I assume, belong to residents racing to save their homes and belongings. But why are people laboring up on bicycles? Why are others strolling along in shorts and T-shirts, some with cameras dangling, like tourists? And who are these teenagers with and without skateboards, trucking along as if to a party?

Next door to me in a house tucked down near the creek, the shades are drawn across the glass-paneled roof, and I assume the people are out of town. But things are serious, aren't they? I rush down and bang on the door. My neighbor Kate opens a slider in her flannel nightgown, rubbing her eyes. She and Bill have been napping with their baby. "There's a big fire!" I announce. "You have to get out!"

"Oh, thank you very much, Jeremy," says Kate, with impeccable politeness. "It certainly was thoughtful of you to come down and let us know."

"That's all right," I say. But does she get it? Bill rocks up in jockey shorts behind her, like a sleepwalker, and she turns and screams in his face. "There's a fire! What the hell do we do!"

I find myself in my kitchen, staring into the refrigerator. What delicacy to save from here? The phone rings. My friend Charlie greets me with estian forced cheer. "How are ya, big guy! Hey, we heard there was a little action in your area. Thought maybe we'd drop up, see if there's anything we could do." "No time!" I tell him. "Well," he offers brightly, "don't forget to turn off the gas."

As I ramble out in search of the gas meter, muffled explosions come from above me with the rhythm of popping popcorn. I find the meter, but don't know which knob to turn. I run down the street, looking for a wrench to borrow. Helicopters and other craft zoom beneath churning clouds, some dumping fluid.

My neighbor next door to George has always been sullen, embarrassed that his big Husky, a secret pussycat with humans, tries as a matter of principle to kill every dog that passes, including my own dear Violet. Now the Husky's owner smiles as he hands me a socket wrench. "Helluva thing, huh?" An old man I have never seen before comes along to show me what to do.

We scramble down towards the creek, clinging to high grass and shrubs, and manage to brace ourselves in position to get the wrench on what he says is the shut-off valve. The old man is happily musing at how imperfectly people understand their meters. He has spent the last hour short-stopping house explosions, having the greatest day since his retirement. Our only trouble is, the wrench is too slight to gain purchase on the bolt. We squander precious time stripping the bolt-head, until my neighbor Bill appears—up and about now, a look of auto mechanic's boredom on his face—wrenches my gas off and climbs back up to his driveway.

A cool press of priorities descends on me. I find myself yanking at my computer components, heedless of my sore back. The machines resist with mechanical obduracy, refuse to unbuckle cables and plugs. I manage to extract my hard disk—eight years work!—and grab a pile of books I wrote, now out of print. I cleverly, symbolically throw some papers in a wastebasket to carry out. I pile this stuff in the backseat of my car, and my trembling dog Violet jumps in on top and begins trampling to find secure footing.

A little red car zips to a stop in front. Margaret O'Neill, a Korean biology student who is my part-time gardener, jumps out with two flat trays of flower slips. "Do I have to garden today?" she shouts. —"Grab stuff from the house!" I shout back. —"But I can see it, Jeremy."

Yes, the fire has topped the ridge and is eating its way down the far slope, less than a mile away.

Already it has devoured a big new unsold house so high on the upslope it has its own elevator. The fire sweeps down fast as rain. "Let's get out of here!" says Margaret, but I insist we have time to fill our cars.

Moving efficiently at last, I grab photo albums, my rolodex and—in a flash of realism—the file of receipts from all the stuff I've bought since the last fire.

"Can't you see it!" Margaret screams. "We've really really really got to leave!" She heaves in the flats of seedlings, slams the hatchback and takes off. I notice most traffic is moving downhill now, though some cars still wind their way up, along with, to my amazement, more tourists on foot.

I start down in my own car, but brake at the sight of Sherri, George's tough blonde wife, who stands in the street oblivious to traffic, staring with glazed intensity down at her house. Sherri, too, works at IBM, an expert in number-crunching software.

Putting their two-year-old on a childcare schedule, she and George saved enough to build on a playroom and an extra bedroom, replace a driveway undone by shifting earth and create a huge sundeck with a playhouse. With children from previous marriages and an au pair arriving, they'll have seven under their roof. Sherri drew up the plans herself, and the contractors raced—extending the Berkeley four-day construction week—to finish before Sherri's next baby came. Their house and garden, the best kept in the area, are the repository of every dollar they earned or borrowed, their own monument to the enduring American dream. And now, here stands Sherri, nine months pregnant, in a stupor.

I buzz down my window. "Leave, goddammit, Sherri! Take care of yourself! Go, go, get out of here!" Sherri doesn't turn, blink or quiver.

In two downhill blocks, I enter a flatter, more suburban neighborhood in the foothills—and find half the population on a Sunday stroll, even as cars bearing wild-eyed refugees roll by. One man washes his car, now and then glancing at the churning sky as one might consult a wristwatch. An economist walks downhill pulling a child's wagon full of books—his idea of evacuation? It seems most people would prefer death without warning: if we see it on the next block, we don't believe it.

I pull over at the sight of a dermatologist friend in tennis shorts, walking his dog uphill. I shout at him, too, to his surprise. Can it be I'm over-reacting? He explains he's merely going to pick up his young sons, who've biked on ahead.

From my car phone I call my tenant, who has stayed the night at her boyfriend's. Oh God, she cries, my cat is locked in the apartment!

Well, perhaps I left a few minutes sooner than I might have on account of Margaret. It's been a long time since I stopped thinking. I pull around the area and come up from the west. The entire mountainside behind the Claremont Hotel is broiling in flames. The hotel itself, a sprawling wood Victorian, will soon be gone at this rate. I park on a side street, so as not to get my car trapped, and head in on foot with my reluctant dog cringing at my heels. A jogger or two still passes. I decide to thumb a ride.

A van speeds up, its side panel wide open, three Latino men in T-shirts and headbands inside. Could they be looters? Ashamed but not unconvinced, I jump in with my dog. The three men are covered with soot, tell me they've been fighting the fire for hours. Driving against the grain with abandon, we scream around the curves uphill.

Sparks have caught a house at the bottom of my block. Our open van charges through a spray of fire. The scene in front of my house has changed, as if an implacable child were wielding a cosmic marking-pencil. The air is black and charged with falling embers. The fire is fifty yards up the street, where I see my first fire engine. Fire is sweeping around on both sides. Eucalyptus trees down by the creek explode like firecrackers. The pregnant Sherri is gone, but her husband George remains on top of his house spraying, his '69 Mercedes in the drive with its motor running. A shirtless young Asian I have never seen stands with the hose on top of Mike and Kate's house. The three Latinos jump out of the van; one grabs my hose, begins watering everything in sight. I get inside the door and grab my basket of keys. The siding on my house is hot to the touch. As I clatter down the side stairs, my lungs are scalding, hot embers fall on my face and hair. Do I mean to die for a cat? My hands shake as I fumble with the heated keys, trying the wrong ones.

I throw open the door: my in-law apartment is the same as ever, unreal that it will soon not exist.

Luke the cat has backed himself into a corner. I grab him and run, feeling careless that I've left the keys behind. Luke is thrashing, he's scratched long wounds in my arms. The fire is one house away. The Latino guys are yelling at everyone to get in the van, wildly begging people to leave with them, but no one will. Unbelievably, a young photographer in fatigues ambles up the road with a huge camera, raising it to snap pictures of a blazing tree, then walking on under the tree and out of sight into the smoke. I'm sure we'll find him dead. The cat fights as if I'm death itself, twists free, ricochets off the wall of the van and out the open side, streaking into the brush and gone. The van is moving now. We plough through falling burning branches, showers of sparks from the wires and sheets of flame until we are back behind the concrete divider on Tunnel Road.

I shake hands with the three guys and thank them, say I'll never forget. One of them turns out to be a doctor. I pry my terrified dog from the corner where she huddles, and the van whips around and charges back toward the fire—which leaps Tunnel Road and hisses from three sides as I reach my own car. Me, I'm satisfied. It's all right to leave now. Maybe 45 minutes have passed since I first smelled smoke.

We take refuge at a friend's house, but the police come by and order everyone out, and this house, too, goes up in the middle of the night. We have ten minutes to pack two family cars. I move steadily with their belongings, depleted, feeling little, unable to stop, oddly aroused. I drive to a house in Berkeley where Margaret has been before me and unloaded my things. I see the flats of seedlings on the grass and notice each flower is singed.

The yard here is full of dogs. Their owners cluster on a balcony watching the huge plume of smoke billowing over Oakland. Will the fire come this way? These friends, too, have packed their cars, and Margaret has fled as if her life depended on it, though the U.C. campus lies between this part of town and the flames. Still, the fire in 1923 burned down on this side a three to five mile strip from Wildcat Canyon to Hearst Avenue.

The children are in another room, sitting hushed as they watch the 49ers game. The 49ers are winning. The camera shows soot falling in the stands, swings up to show the pillar of fire thirty miles away.

In another room we watch fire coverage, changing stations away from each pompous interview. The camera crews by now are up in the hills. We see footage from hours ago showing an apartment complex disintegrating. The condos in Hiller Highlands went right away, like cardboard boxes, and people fried in their cars when the fire raced into a bottleneck on a narrow road called Charing Cross.

One man in the room, a corporate lawyer, has had to evacuate his million-dollar home with his wife, two children, three stepchildren and many pets, and spends the afternoon whining how he will sue city officials and the fire department. He takes his family to the Fairmont that night, feeling his insurance company owes it to him. The next day he goes back to sift through the ashes and finds his house intact.

As the afternoon moves on we see the fire veer South away from us, crossing freeways and threatening to sweep down into the densest part of Oakland. The TV tells us the governor has flown over the fire area. What a relief! There is more talking, packing, driving, and somehow it is late before I go to bed with my dog in a spare bedroom of a house where I used to live ten years ago with my love Melanie, an ex-Mormon from Utah set loose by the Sixties, and Melanie's children. In the ungluing of our household, the collapse of respectable dreams, we have become known quantities to one another. It seems natural that I take refuge as a brother, ready in the night if need be to help her evacuate a house we nervously, excitedly moved into as lovers.

How odd that this house is standing, that the little girls are not there, they have grown and gone off to college.

Violet the dog has bad dreams and yelps pathetically in her sleep.

For some weeks I continue in my manic phase. Days are spent running from one place to another, returning phone calls, trying to replace papers, filing insurance claims, looking for housing, beginning to buy back possessions. We learn that the fire destroyed 3354 homes and 456 apartments, killing 25 and injuring 150. Each night there are community meetings. All spare time is taken up with fire stories, told urgently as if the listener will not believe. And every day I am bubbling with

good cheer, except for moments when I am overwhelmed with sudden sadness, and promptly collapse into sleep.

I don't know why I have yet to hear of a homeowner who died on his roof, hose in hand. George and Sherri watered till the last second, unable to concede, and got out without saving a thing. I later learn that she heard me shouting as if from far away, but could not take her eyes off her house, even as every three minutes she was wrenched with the pain of a contraction. She went from the fire to a hospital, where they told her she wasn't ready yet. Sherri and George spent that night lying fully clothed on a hotel bed, unable to sleep, move or speak. The next day a healthy baby girl was born.

But there were those who went down with the ship. An 85-year-old woman near me made it her job to garden the slope of vacant lots. She wanted the whole block to look good. They say she refused to leave when the fire came her way. A patrolman who had volunteered for a Sunday anti-burglary shift died trying to lead out a group of people on foot. A battalion fire chief jumped from a retreating fire truck and went back into the flames. He had just pulled out a woman overcome by smoke, when a power line fell on him, shattering his helmet and killing him instantly, along with the victim he had nearly saved. A number of innocents were found in basements with wet handkerchiefs to their noses. They died from suffocation when the super-heated fire sucked out their oxygen in one gulp.

The most poignant story involved a young couple on Buckingham Drive, on the slope above Charing Cross where the fire first erupted with such ferocity that they had less than a minute to get out. The woman ran down the drive, jumped in the first car and headed out. The man dashed back, grabbed a guitar and sped downhill in the second car. He hoped to catch up with his wife, but the driving was tricky. He went to his in-laws' house in Lafayette, going the long way north through Concord since the tunnel was closed and hundreds of cars had been abandoned on Route 24. But his wife wasn't at her parents' waiting for him. She had turned uphill instead of down, and was probably dead by the time he got out the driveway.

People surprised themselves with their own survival. Some non-aerobic types jumped from jammed-in cars and ran down through burning fields the whole way. On the fringes, houses were saved by never-say-die owners on roofs with hoses—or by strangers, like the men in the van, passing through the fire area after everyone had fled or been ordered out. Police ordered some people to evacuate so fast they could not bring anything—only to find that neighbors returning from trips on Sunday night were able to drive in unimpeded, load up their possessions and even, in favored locations, save their houses while others around them burned.

It turns out an angel watched over the photographer I saw. He was saved up the hill by a software designer whose house blew up as he stood on the porch watering it. Finding himself too hefty to run and manage his four dogs at the same time, he grabbed a plastic child's wading-pool, dragged it to the middle of the street, filled it with his still-functioning hose and sat in it with his dogs huddled around him. When the photographer came by, the designer pulled him in. They crouched together, the photographer shielding his camera as waves of flame swept overhead. The photographer was elated by the excitement of his first assignment. When the heat grew too intense, the computer guy bolted, running his dogs through the flames and uphill, into burnt-out space. He looked around thinking the photographer was on his heels, but the insouciant lad had disappeared, following the fire down, snapping pictures all the way.

Most of the locked-in pets mercifully died of suffocation, but what about the ones outside, like Luke? Among the emergency numbers victims could call was a lost and found animal exchange, which brought about hundreds of reunions. The favorite story of the hills is the cat who had disappeared four years ago, then returned to the site of his burned home. But so far, Luke has not appeared.

Spirited mutual-aid groups have sprung up to press city government for cooperation. At meetings I see half the survivors sitting stunned and drained; the other half speak with wild energy about immediate rebuilding, which for many is the only way to reclaim full "replacement value" insurance.

The spontaneous energy is amazing. A builder's wife, a month after a miscarriage, got news of the fire while attending her father's funeral in Fresno.

Being away, she and her husband lost everything, their own house plus two uninsured homes they were re-building and owed money on. They camped in a relative's basement with their three small children. But she and he are the most relentlessly cheerful people in town. Four days after the fire, she had engaged experts to replant her area with fire-resistant, water-holding plants and trees.

Houses will rise again in the Claremont Canyon, and one of these days, will burn again.

A problem for us manics is the stream of phone calls from everyone we ever knew, tracking us down to wail how unhappy we must be. I'm surprised to hear from my ex-wife as well as many women I assumed had forgotten me. It's peculiar how many feel guilty. A woman friend of mine cannot sleep until she makes sure the house of a now-hated former lover and his new wife has survived. Another cries to me in anguish, "Are you telling me there's absolutely nothing I can do?" "Marygold," I say, "this fire is not a relationship between you and me."

Those whose houses survived were often in worse shape than those who were wiped out. The most desolate man I encounter the day after the fire is a big discount store owner wandering the two blocks of houses which remain at the bottom of the hill, cradling his infant son, haggard and unable to speak. His problem is that the firemen drew a defense line above the Claremont Hotel. His big stone house just beyond the edge was saved, while two neighboring houses inside the line burned. His friends from these houses—one of whom lost two years' work on a book she was writing—are chattering away with animation, not noticing his silence.

The dermatologist comes along with his dog. Turns out his sons did bike back before the flames got them, just as he had assumed. The three of them were heroes, picking up an unmanned fire hose to help turn the blaze.

Every hour or so I am devastated by the pang of something left behind. I saved my snapshots but not the Super-8 film of my two sons as children— touching scenes which I can play in my imagination, though I have not, I remind myself, seen the footage in 18 years. Why didn't I think to copy this film onto VHS, as well as the old reels of my brother and me as kids, that were mistakenly left in my care 15 years ago? I could have grabbed all these and thrown them in the car in half the time it took to have the fax machine fall apart on me.

In the '89 fire I lost half my books, and afterwards wiped soot from fifty irreplaceable volumes with a soft rag and benzine. For the past two years, I've never known if I had a given book or not. But I had learned I could live without them, and a good thing, too—because now I have no books, records, furniture or clothes. Fifty times I passed the front closet, where three leather jackets hung—and everyone wants to know if I saved my Oscar. I lost my electric bed, which never did help my back but caused interesting problems for anyone in bed with me. I lost all the personal letters anyone ever wrote me—when it would have been so easy to pick out the ones close to my heart and put them in my safe deposit box. I lost the old records I had lovingly taken from my mother's apartment in Indianapolis when she died, including "The Ballad for Americans," the '56 campaign record making fun of Ike and Joe McCarthy; Elvis singing "Heartbreak Hotel"; and a scratchy 45 of my father at Ft. Davis, North Carolina in 1943, about to be shipped out and singing, "Take Good Care of Yourself." I also lost, among other birth, death and marriage mementos, a tiny initialed baby ring given to my first son when he was born—a treasure I've been holding for him ever since—

though he has never seen it and would not know what to do with it.

In my manic mood I felt relieved because I didn't have to carry this baggage through life anymore.

The most symbolic loss was 150 notebooks I'd kept since 1962, recording odd thoughts, events, writing ideas. Now and then I really did look back at these to remind myself of what happened. But they embarrassed me—the tone was too youthfully judgmental to suit the person I had become. Realizing that some comments could burden my survivors, I had willed that these notebooks be burned unseen after my death. Now the fire had done it for me. No need anymore to torment myself with the thought I should review them and save the possibly valuable passages. Nor do I have to worry about the manuscripts of two novels I held for years awaiting one more draft.

In the days that pass I find myself instinctively heading to my non-existent home each time I get in the car. The one morning I really do go back, I drive up Alvarado Road, which for the first two curves is so exactly as usual that it is an utter shock when I turn a corner and hit the vast black slopes of silvery white ash where once I lived and walked, smelled the flowers and spotted graceful deer. "You mean it's not a dream?" I ask the cop who checks my ID at the barricade.

Here and there chimneys stand, sometimes with two or three fireplaces hanging suspended one over the other. One sees the occasional concrete garden wall, with metal chairs and tables poised for a tea-party. But along my block of Grand View, there is nothing. The creek walls acted as a fire chute, pushing a wedge of inferno downhill at 2000 degrees Fahrenheit. The burnt steel skeletons of cars hang upside down over ditches and canyons. At first I can't find my house. All that remains is five feet of concrete driveway at the edge of a black hole. No, there is the little wall of stones, piled one on top of another without mortar, which divided my lot from the street. It seems every day I replaced a toppled stone; now this wall is the only structure remaining. "Good fences make good neighbors," I think idiotically.

The lot itself is a pile of ashes. Down in the still-running creek I see bathtubs and twisted blackened hulks that must once have been a refrigerator and stove. My house and possessions have been reduced to rubble, no speck larger than a burnt cereal flake. But downslope I recognize what was once my old filing cabinet, which stayed with me East coast and West for 30 years, and which held old writings I thought had surely burned. I wade down through knee-deep ashes, grabbing metal rods protruding from foundation strips to keep myself from tumbling 100 feet to the bottom. The rods bend like dandelion stems and loosen from the crumbling concrete. I have to keep moving or get spilled. Finally I brace myself against the

blackened metal of the cabinet. I stick my hands into a crack, trying to widen it.

Inside, I see a miracle—a sheaf of papers. I see letters, print—the lost manuscripts! I strain against the metal till I can wedge my hand inside. And the pages turn to dust in my fingers.

After a moment, the terror drains from my body, and I see the immense devastation has a kind of wild, transcendent beauty. Beneath me, delicate yellow fencing is webbed along the hillside to hold back the earth. Beyond, the sepia-toned landscape opens as far as the eye can see, blackened trees standing like primitive totems in fields of snowy ash. Hearths are scattered on receding slopes like altars strewn with collages of sacrifice to the arbitrary violence of the gods. Blankets of sprayed grass seed shimmer in brilliant astro-green in the canyon depths, and tiny flowers already poke their heads through the mud along the creek.

"Death is the mother of beauty," said Wallace

Stevens, in his poem *Sunday Morning,* which might apply as well to the Sunday morning of October 20.

Do I feel this all the time? No. Would I feel it if I hadn't rescued my dog and my disk, if, say, I had been like the painter in his 60's getting ready for his first one-man show, who had every painting turn to smoke?

But everything we hold onto will go anyway—a fact that defies the unthinking assumption of daily life. We live, we must live, mostly as if we don't believe this. The difference with my house and possessions is that they went all at once, with me there to see them going, and in the shock of this I am consoled by intimations of order and relief. For years the voice of Billie Holliday has been too painful for me to bear; now I listen again and am thrilled.

As I walk down at dusk from the ashes of my house, I pass a new Infiniti pulled over to the side of the road. A lonely rich guy, having finished his day at the office, plods in deserted shadows with an open can of liver scraps, softly calling the name of his lost cat. Cooing cries of pet owners rise in the murky stench, echo along purple canyons gleaming brighter than ever in the polluted gorgeous sinking of the sun.

Driving Home

Marion Abbott Bundy

'm racing north on 13, having spent the late morning in the library at Mills College. Ironically, as I will reflect later, I'm in the midst of reading D.M. Thomas' novel *The White Hotel*. I've just finished the section that describes in hideous detail a devastating fire that kills hundreds of guests of a white resort hotel. I'm feeling shaky and spooked.

But there's no time to ponder Holocaust metaphors now; I've got to shift gears and get home in time for the Emerson School Walk-a-Thon which begins at one o'clock. As usual, I've cut it close, leaving campus at 12:40. Thank God there's not much traffic.

Indian summer—but so blustery.

Earlier, sitting on a bench waiting for the library to open, I watched eucalyptus leaves chase each other in circles like frenzied puppies pursuing their tails. Heard strong branches crack with the force of the eerie easterly wind.

Is that fog ahead? May cool things off.

No matter how harried I am, the drive along 13 with its serene leafy backdrop is always soothing. Dense trees form a spectrum of greens, from gray-green eucalyptus to forest-green redwood to black-green pine. The tree trunks, dead leaves and dry summer grass make a band of earth tones, while bursts of red from bottlebrush and pyracantha dot the palette. I wish I could paint.

I whiz along, imagining how, from a distance, my orange Volvo looks like a hot tangerine comet streaking through a cool verdant sky, like a Clyfford Still abstract. I wonder how these trees withstand the extremes of freeze and drought, the ongoing assaults of smog and pollution.

A forceful gust of wind shoves my car sideways into the next lane. I correct the steering and crank up the window against the roar of the road and the swirling debris. The hot wind is charged; it gives me a chill.

Earthquake weather—and I'm riding directly on the Hayward Fault. I floor the accelerator and speed past the gold-spired Mormon temple that is more Oz than Oakland.

It's not fog at all, but a sinister charcoal cloud charging across the almost white midday sky. A pregnant thundercloud, the kind you see up in the Sierras—or Kansas—but not down here. Maybe this odd wind is ushering in a storm; we're desperate for rain.

Suddenly the cars ahead of me go haywire, scattering like cockroaches surprised by light. They swerve off to the shoulders. Some stop dead in the middle, yellow lights flashing. Those with a clear shot, exit. Others back up to reach the off ramp. I downshift, pump the brake, jerk my head around to make sure I can execute the maneuver and slip across the two lanes to safety.

Or so I believe in the instant before I punch on the radio to learn of a six-alarm fire sweeping through the hills, raging out of control, fueled by this insane wind.

"Preliminary reports say the fire in the East Bay is centered above the Caldecott Tunnel. It's jumped from Highway 24 to 13 and is advancing on the Montclair district of Oakland and over the hill into Berkeley . . ."

The thundercloud is actually noxious smoke pouring into the sky with almost reverse gravitational force.

I zigzag through unfamiliar streets, up the gradual incline, my eyes suddenly transfixed by a brilliant spectacle in the distance—a monstrous dancing sheet of red and orange fury. Groups are gathered on corners, gaping, silent. The blaze is no more than a mile from where they stand.

"That fire's picking up speed, they just can't seem to get it under control yet. Crews from three counties have been called to the scene . . ."

Now there's traffic, and it crawls. I pound the wheel, lean on the horn. "Pay attention!" I yell at no one and everyone—at myself as I tap the car in front of me, jam on the brake, stall. "Shit!"

" . . . massive fire still tearing through the East Bay hills, fast approaching the landmark Claremont Hotel . . ."

If the Claremont goes, so will we. Our house isn't three hundred yards—the sycamores—kindling.

I flash on the fact that I wasn't home when the earthquake hit, either. Why does natural disaster strike the minute I leave the house?

Finally a break in the line of cars on College. I scoot across and barrel up my street. Past my daughter's school, where a crowd on the playground stares up.

"If you can feel heat or smell smoke you should evacuate . . ."

Home. It's still here. I slam into the curb, yank off my seat belt, leap out of the car. Palpably hotter here—a blast furnace. Smells like someone forgot to open the flue.

The kids are wandering around the house. The radio is blaring, but I can't make out any of the words. Stephen and I disagree on the degree of danger. We have words. I want to flee immediately. He wants to wait and see.

I start packing. What to take? Where to start? I float from room to room with a shopping bag, unable to decide what to put in it.

The phone rings and rings; I can't answer it.

Within minutes Stephen and I load the wayback of the station wagon. I've focused enough to fill the laundry basket with picture albums and my rolodex. The kids, at their father's urging, have each chosen something special from their rooms. Will clasps his stuffed tiger; Emma her doll. She's more concerned with her loose tooth than her house possibly burning down.

Before we pull away, just like always, I ask the children if they are buckled up. They click their seat belts into place. Our nervous little dog hovers between them, panting, nose in the air, ears cocked; she's on full canine alert. I take what may be a last look at our house . . . and point the car down the hill.

We're together. We're safe. The tooth fairy will find us wherever we are, I tell Emma.

from Burn, Baby, Burn

Frances Rowe

 couple in their forties stood clutching each other on the fringe of the crowd. They were focused on a point in the inferno which was directly in our line of view. The flames licked and roared at some houses not yet on fire but oddly seemed to veer away from others. Trying to predict what would explode next was like trying to predict the path of a tornado. Suddenly the couple both yelled out.

"Oh God, that's our house!"

"Yes, it's our house."

"Oh God, no!"

They turned to us as if to appeal for help, but we were frozen in place in spite of the heat.

The woman was wailing now, and tears drenched the man's face. He called out a name that wasn't clear, over and over. A pet? A pet-name? Surely not a person? The next time he groaned, "Oh God, no," we joined in involuntarily, and it became a dirge, chanted by the Greek chorus of spectators. Then a man and a woman from the crowd moved in and made a sort of ring around the couple with their arms. Now that the paralysis of horror was broken, others touched the hands and pressed the shoulders of the agonized pair.

"We're sorry, so very, very sorry." ☐

What We Take And What We Leave Behind

Bernadette Vaughan

She feels suffocated, as if her ribs have collapsed onto her lungs. It's a familiar sensation: the panic that visits from deep night, wakes her in the dark and draws her out to pace the floors, fighting for control, fighting to breathe. For years she has struggled to track it, to give it a shape, or a time or a location. Sometimes it seems as though she finally has it, but eventually it eludes her; glides away and returns, silent and looming, to press dark formless hands around her windpipe and pile muffled stones on her chest to entomb her.

Now, though, there's no mystery. The air is urgent. Energy rushes the windowpanes, somersaults back and forth and fills the drought-parched canyon, rattling the trees and agitating the sere brown grasses. Charred particles whirl and drift, blown by the wind that drew its first breath in the night and hyperventilated the dawn. Earlier in the day, the Redwood Park dogs and their walkers uneasily lifted their heads, sensing and sniffing the disaster that is happening now, in the afternoon. The air smells of smoke and chemicals, but high over her house the October sky is still blue, still innocent, denying all involvement.

Suspended above her head, hanging from an almost transparent length of fishing twine, the vanes of the radiometer, like little wings, respond to the energy, to the air. Her son, who is missing in the fire, gave it to her as a birthday gift in April, reflecting his delight in the scientific discipline he has embraced. On this October afternoon, she knows why she is having trouble breathing. The television commentators have reported that the fire is out of control. They relay the entreaties of the fire department: all hill dwellers, please, please, evacuate. Abandon your homes. A worried friend who lives downtown, near the lake, has telephoned to say that the fire is moving like a racehorse across the hills; it's time to think about leaving. Yes, she says obediently, Yes. I'll be ready in half an hour. She begins to move deliberately through her house, begins the wearing process of packing: first to the woodshed, to find boxes; retrieving newspapers, for wrapping fragile objects, from the recycling pile;

making lists; breathing, in and out. She knows she must stay conscious, concentrate on a step by step progression of departure and wait for the telephone to ring. Earlier, when the phone rang, she leapt for it, into the air, across the room; everything in her, pulse, brain, all her vital signs, leapt into the newly redeemed and ordered present; the bright, irrefutable now telescoped into his voice, about to tell her that he was all right, that all was well, in spite of his apartment building burning, as the television showed, burning to the ground, exploding, being swallowed up; the structure, a wall of flame, the trees, vertical torches; some tenants dying, others, watched by video cameras, running barefoot for their lives through the blazing trees; that despite all this, he was safe, he was whole, unscarred and unconsumed. But it was not her son on the phone; it was her concerned friend, the one who was coming with a truck to help her move.

Perhaps he was already in the hospital.

Moving about her study, sorting papers and floppy disks and photographs and silly mementos, mentally palpating the terror she always feels when someone says "burn unit," she allows herself to watch him lying motionless, wrapped in bandages, threaded with transparent tubes, a comet at the end of its trajectory, Icarus brought to earth, hurt, perhaps badly, but alive. She feels hysterically relieved. He is too beautiful to burn, she tells herself, he will not have lost his features, his hands, his wonderful eyes full of sparkling shadows, all his amazing beauty.

Memory kicks like a fetus: she remembers Teresa, whom she first saw the day she walked into kindergarten, her first day at an English school at the outbreak of the war. Teresa's face was burned, the right side of it, including her eye, which was seared and bloodshot and often suppurated, oozing yellow pus. The other was lustrous and clear, fringed with long tangled lashes, glowing and expressive, the color of a summer river, incandescent. Teresa had fallen into the fire; her mother had not been able to save her. She was scarred for life, one side of her face untouched and delicately tinted, rose and olive, the other crumpled and scorched, looking as though the skin in places was pasted on, smooth and taut, where the flesh had melted like wax in the fire and then hardened into something else, defensive and alien, a carapace of celluloid or shiny white paper, nothing like skin at all.

The beautiful also burn. She hears the uncompromising reminder, and into the box she has placed on the floor, she carefully stacks the essentials, which will go down to her friend's house by the lake for safe keeping: the last will and testament, the IRS documents, the insurance premiums, the unfinished manuscripts, the beloved, irreplaceable books, the framed poster of the children, just past babyhood, dancing about the tiny living room of the house in Woodstock. Their clothes, she sees with surprise, are easily a year outgrown. Why has she never noticed that before? Jabbing, the familiar regret at her failings as a parent moves up from a hot red waiting place inside her and takes its place beside the dread of death by burning.

The phone rings again. Now it is her daughter, wanting to know if she has heard from him, wanting to know her plans. His line, they tell each other, is busy, but that's because the apartment building burned up within the first hour of the fire; it was the first to go. Soon, to avoid more confusion, all the telephone lines will be down, the way they were after the earthquake. That's the way these things are handled. They agree that it's unfathomable, but doubtless contains its own logic. "You've got to get out of there," says her daughter. "If the wind changes direction the flames will go up the hill to Montclair. You'll be trapped."

The voice on the line is strained and high, frightened backwards into childhood. She reassures her, tells her that friends are coming to help her.

"Where will you go?" says her daughter.

She doesn't know what to answer, because she doesn't want to go anywhere until she knows he's safe.

"I've got the dog. That's a problem, because just about everyone I know has cats. But I don't think I'll go to a shelter," she tells her daughter. "Can you imagine? I mean, this enormous creature, in a shelter? But there's that motel in Pacific Grove, you remember, that takes dogs. I'll go there. Don't worry, darling. We'll be fine."

She knows she will not go there. She will pack the boxes and load her car but she does not think she will leave the house, the womb with its electronic umbilical cord, the telephone.

Carefully, forcing herself to look at them, she

packs the framed photographs from her desk and bookshelf. Her son and her brother share a frame; she keeps them in proximity because they share an ineffable vitality, which she has always supposed is what people mean when they speak of "family likeness." What she thinks, but has never explained, is that she sees their souls stitched together, flying like twin kites on a single string, their radiance streaming against the sky. Her son as a little boy, one of the most appealing children she has ever seen. Always uncompromising, always challenging her. She remembers how peaceful she always felt when he acted in his own best interest. Just look at what he did to that gorgeous John John Kennedy haircut she paid a fortune for in New York, just before she left her husband. Two years old and determined to have the last word, to maintain control of his life at least through his appearance, he had run to her bedroom, taken her nail scissors out of her drawer and snipped off two chunks of hair, close to his scalp, one above each temple.

She remembers him shortly afterward, striding into the room like Mussolini to interrupt a bitter group dispute, one of those endless sixties disputes about fidelity and freedom that always, in retrospect, preceded the breakup of a household. "Grownups!" he had roared. "I've had enough of all this quarreling!" They stopped in mid-sentence to stare at him, uncomfortably aware of their self-absorbed absurdity, the cruel, reversed truth of the situation they had made for themselves, for the children. Ragheaded, grimy-cheeked, enchanting, he smiles confidently from the frame.

Alongside this snapshot, one of her brother, debonair in the dress whites that befitted his membership, exalted beyond his age by the exigencies of war, in the Senior Service. The picture was taken in Egypt; he had just been made the youngest Lieutenant Commander in the British

Navy. Men's lives hung from his judgement, from his handsome, sunburned authority. And all that potency evaporated in a single, irrevocable moment of error when the mountain loomed in front of him, and after so many narrow escapes during the war, luck and time turned traitor and ran out, reducing his fragile flying machine to ashes on the ground. Now, far away, the objects that made him a legend of excellence, the things he competed so fiercely to win, to use as barter for love—the demanding silver cups awarded, first at school, then at University, then as a Midshipman, for running, for boxing, for cricket and Rugby—remain in the world and claim dominion. "Oh death, where is thy sting-a-ling-a-ling? Where grave thy victory?" They sang that song, she remembers, all those bright boys, barely more than children, at an age where in a sane world their mothers would still be making appointments for them to see their pediatricians, in the cockpits of tiny fighter planes, pitching out headfirst into flak and glare and tracer bullets to die over Norway or Berlin or Arnhem. Now her son is older by three years than her brother was then.

She puts the containers of snapshots and memorabilia—she has a file for each of her children: artwork, notes, teachers' commentaries, letters—into the box. If he's dead, she thinks, there might come a time in my life when I can bear to look at them. Was she, she wondered, so different? Women lose their sons all the time. Young men die in droves, drop like snowflakes: there are wars, automobile accidents, inner city violence, drugs, and testosterone sprayed like graffiti, like Uzi fire. Life goes on. There are mothers all over the world walking around wearing their viscera on the outside; their lives are blitzed and life goes on. But does it? Does it? Or do they die too, inside? Is the world full of women who are in fact the walking dead? Does anybody hear them sobbing in the night? What is meant by "adjustment" and "coming to terms"? Women die on their husbands' funeral pyres in India; what would happen to the world if women died on their sons' graves? What would the status quo do then, poor thing? Hide its head under its wing, poor thing.

My mother went mad, she thinks. That's how she dealt with her son's death: she went finally, irrevocably, over the edge and into utter chaos, taking to her bed, her face turned to the wall, the day she returned from identifying the charred, broken body of her first-born, recovered from the site of the airplane crash at the foot of the snow-covered mountain in Wales. Her mother walked back into the house, up the stairs and into her bedroom; she went to bed and stared at the wall, dry-eyed, refusing all food and rejecting all solace. Her mother, always so verbal, so colorfully loquacious, was speechless for a long time. Then one day she said: "I held my son's heart in my hand." She said: "It was burned dry, it was like a lump of soot and there was a big hole in it."

Will that be what I have to do, she thinks, will that be what is required of me, if my son is dead in this fire, to find his heart, find that something has eaten a hole in his heart, that some old desolation, from babyhood or boyhood or adolescence, has hollowed out a dark place in his heart and died there? Will I be expected to live with that, cook and eat and work and interact, to adjust and come to terms with that knowledge?

There's a name for it: Post Traumatic Stress Disorder. That's for men who survive war. But what's it called for the bereaved women, the mothers and daughters and little sisters with their psyches in bodybags, nerves and hormones and biochemical connections hacked off and left dangling, like sabotaged lines of communication? They drift, these women—distracted, redundant, jobless—trying to remember that those terse official telegrams are not legal tender, will not buy food and lodging and education for the remaining children, that life goes on. History is full of them; the world is full of them and nothing grinds to a halt.

It's one of the great mysteries, she thinks. She finishes packing her papers and goes back upstairs. Now she has to sort and pack the domestic treasures, precious and functional things that should not be irresponsibly consigned to smoke and looters. The Limoges dessert plates with the matching cups and saucers; pictures; some artwork. Some linens perhaps. Anyway, restoratives, things that ease the day's beginning, reassure its end and in between burnish the body of quotidian living— provided of course that you are one of the lucky ones, one of those who has not been eviscerated by irreparable loss. First she packs a Safeway grocery bag for her dog: leash and bowl and blanket; large Thermos jug of water; kibbles and Brewer's yeast

and the careful, loving medications, the pharmacopoeia of pills and ointments required by her elderly, arthritic child, the four-footed household member. The dog is restless, but trusting; padding up and down the stairs, returning to her bed, circling it, treading it down, collapsing with her usual grunt of effort, then lurching up and worriedly starting the pattern all over again.

Effortlessly and unreservedly including animals in his perception of reality, he is the person the dog loves best in all the world. Once, in a fit of hygienic zeal, she bought poison to set out in the basement for some invasive field mice. He found it and confronted her, explaining to her why she could not do this, how if she did, she would change in a terrible way, forever, and never be herself again. She had thrown the deadly stuff away, wrapping it carefully so that no little unsuspecting creature could swallow it. Knowing that the dog is reflecting her present disquiet about him, she retrieves her steps, back down the stairs to the first landing. She stoops to rub the dog's head and murmur a message of comfort, soothing them both for a little while.

She works quickly, amazed to discover how easy it is to make choices: what to take, what to leave behind. A couple of pictures, a Spanish mirror, the painted, carved Haitian owl. The woven Guatemalan pillow, the ship in the bottle. Everything slips easily into place, into boxes. She knows that if she leaves the house to the fire, she will have to find time to mourn her home. Otherwise, she knows, she will be prey to depression, or to the tight-lipped craziness worn by people whose faces are still vivid in her mind after all these years, the ones who lost their homes and families, the entire fabric of their lives, in air raids.

You could fall asleep at nightfall and wake a homeless orphan before morning. There was a shelter, dug into the ground at the end of the long garden and mounded over with earth and gooseberry bushes, but it was never used. She had been glad, because sometimes, playing with her friends, they forced the swollen door and peered into it, repelled and frightened. It was dark and damp, smelt like a tomb; it made her shiver.

She remembers how before going to bed she used to leave the essentials—no treasures, no favorite articles of clothing, no books or toys or beloved objects, nothing that might get broken or fall and have to be retrieved, impeding escape—just her school clothes, neatly folded on the chair beside her bed, her homework in her satchel and her gas mask. So when the siren screamed in the dark and ripped her from sleep, she could grab everything in a sweeping armful and run downstairs, her heartbeat and breathing suspended, to crawl onto the mattress her mother kept under the huge dining room table, wrap herself in a blanket and stare at the livid night through windows latticed with brown paper against shrapnel and flying glass, until the droning planes passed and the bombs missed their mark, or fell at random—she always thought Random was a town on the South Coast—and once again they were spared; once again they were not consumed.

On a night like that she saw the stricken airplane, pricked out in black in the frantic white beam of the searchlight, transfixed, like an insect on a pin, just before it burst into flames, spiralling down to earth nonchalantly, as though the pilot had made a choice and was doing a brilliant trick that he'd never tried before but had vowed to do just once before he quit flying. And then the two tiny figures, dropping through the angry sky, blossoming like delicate, transparent mushrooms as their parachutes opened and the night wind blew them gently away, out of sight, into the pocked night, toward the burning buildings or perhaps the sea. They called their planes "kites," she

remembers; the sea was called "the drink"; when they crashed, it was "a prang." Their speech was littered with the argot of war; they spoke, for the most part, not words but incantations, a gallant, echoing language of talismans that made sacred their euphoria and their terror. She remembers how grateful she was when her son finally passed the age at which her brother died, which is to say, when he passed the dangerous draft age; she could breathe again. It was as if a bony index finger, pointing at his face, had been knocked away.

Not, she reminds herself, that it's any guarantee.

She works quickly but thoroughly, often running downstairs to retrieve an overlooked necessity (her castanets, a favorite shade of nail varnish, the essentials for cleaning off her makeup, the new T-strap shoes, the Vitamin B for stress). She is satisfied with her work so far; if she has to leave, she will still have the rudiments of a home. She has packed pillows, the handwoven rug her children gave her for Christmas one year, some towels, a robe, soap, toothpaste.

She wonders what to take from the kitchen. Certainly the trio of strutting steel piglets, which are in fact trivets but which decorate a wall. Three little pigs. She has always liked that story, always identified with the search for the warm safe haven from which one cannot be dislodged.

For her it had been her doll's house, an enchanted habitat where as a child her imagination swam like a fish desperate to dig a nest. Kneeling before it, fervent, adoring, placing the miniature objects and spinning out histories to embellish the wire and wool doll people, she invented games that simultaneously contained, illuminated and transformed her physical surroundings. In the worlds she conjured up and in the texts of the ardent whispered narratives that chronicled them, order, grace and permanence prevailed.

When her brother, an adult while she was still a child, broke off his engagement to the lovely girl from Malta, all the gifts had to be returned, including the minuscule exotic furnishings, the tiny books and pewter plates and hand-sewn bed linen, that the fiancee had given to her, the prospective little sister-in-law.

She had thought she would die of the pain of handing back those infinitesimal objects, so exquisitely invested with desire, that she had thought were hers. Now, fifty years later, she has the exile's need for an enduring hearth; sometimes, when she presses her forehead longingly against the windows of toy stores that display doll houses, she wonders if, in some reality warp or parallel universe, she could find herself inhabiting one. It has occurred to her that this might be all she need know of heaven.

Suddenly, she feels tired and rather scared. Her head is beginning to ache—the slow, deep, dark congestion that heralds a migraine. She feels bereft.

Restlessly, she roams the house, her movements becoming larger, less coherent and purposeful. She has lost interest in packing. She picks things up and puts them down, uncertainly, wondering whether she made a mistake in not including them. The telephone, when she picks up the receiver, gives a dial tone. Her line, it seems, is not down yet; still he hasn't called. The fire is still burning up the hills, and it seems closer now, although the wind is still blowing the flames away from her house, down toward Rockridge. Outside, her neighbors blow leaves from their driveways and drench their roofs with water, activities which to her seem futile, though understandable. One must, it seems, do something. The television chatters on, the commentators looking stern and desperate. They show a mysterious map of a familiar, nearby area. She can't understand it, and decides not to try. There is an announcement: the Oakland Fire Department has decided to make its stand at Broadway and 51st Street. Here the line will be held. She wonders what this means. Have the firemen held council and drawn a line on the ground, as at the Alamo? What good will this do, since the fire has claimed the upper hand? How will they prevent it from consuming Rockridge and downtown Berkeley and Oakland as far as the estuary; how can they do that, when nothing can stop it?

She opens the French doors and walks out onto the balcony to wait for her friends and to watch, over two hillsides and across the hidden freeway, the scarlet flowers blossoming in bunches in the trees, crackling and leaping, insolent bouquets tossed on the wind.

She feels explosions through the soles of her feet, standing there on the balcony: dull, muffled brisances that shake the foundations of the house; and implosions, substance collapsing inward on itself as the broken heart at the center yawns open and the dwelling ceases to hold. She grips the balcony rail and listens, leaning out over the drop to the canyon below. She thinks she can hear the screams and the shouts of the firemen and the hiss of the hoses.

There is a roar overhead; it terrifies her. Looking up, she sees the police helicopters circling an area of new fire, closer than the last. The sky is smoke blackened, torn, showing a red underlining. Some new horror is developing.

She runs indoors, calling the dog. She grabs a blanket that she has folded neatly near the front door, runs back out onto the balcony and drapes it over her shoulders, crouching on the deck, holding on to the dog's rough coat. She stares at the sky and she sees two small airplanes diving and banking on the edges of the fire. One is itself on fire, she is sure of it, one is burning and any minute will go down. Down, down into the deadly, blazing spin, molten metal pure and shining pouring itself in a spiral of death out of the sky and into the earth, the young radiant life in the cockpit suddenly aware that it is over, that he is going down to blistering death. She thinks she sees a tiny black figure, a flyer's figure, perched on the edge of the twisting ribbon of flames, his flying helmet and goggles pushed back; she hears him singing: "The bells of hell go ding-a-ling-a-ling, for you but not for me/For me the angels sing-a-ling-a-ling..."

The voice trails away as the vision vanishes over the ravaged horizon. She leans against the balcony rail, watching the sky, inexpressibly saddened by it all, by the waste and tragedy and futility of it all. Then she remembers the second craft, which may not be lost because earlier it had peeled away from its stricken partner, and she hunches forward, scrutinizing the smoky overcast and beyond. Suddenly it is there, in a patch of clear sky overhead.

It is intact. Observant, steady, its engines throbbing, it hangs high above the fire, and she watches the sunlight glint off its wings, sees it spread-eagled, white against the blue of the October afternoon, as if suspended from a length of fishing twine. It circles slowly, steadily, far above the havoc of the leaping flames, a silver hawk riding a thermal current. She feels her sorrow subside, her hope returning. Below, the fire swallows the hills. She redrapes the blanket across her shoulders, stretches her legs along the deck and eases the dog's weight across her lap. Her breathing is calm. There is still time. She settles in to wait.

BERKELEY-OAKLAND HILLS, OCTOBER, 1991

Fire takes the right-of-way down Tunnel Road
plays basketball with the firemen
leaps higher
throws a net over Grizzly Peak
catches slow squirrels in the lineup
lights eucalyptus candles for its birthday
eats three thousand homes
warms up for the ninety-two Games
learns to throw terrorist bombs
illuminates the page
never stops to reflect
waves to the City from above the Hotel
sticks out its tongue for the cameras
laughs
erases the sun
embraces the lonely
ignites our dreams
grabs the fireman each morning
the rest of his life
he wakes up stamping

—*Sharon Olson*

from RISING

From the bus stop bench
where I watched the smoke
change from grey to black,
where I watched it curve
and reach and steal,
I could not see the house
for the mass of smoke.
She was already hidden,
a great thick blanket
of threatening swirls
was smothering her.
       I had almost wished
       she would just blow up
       right in front of my eyes
       so I could know what it's like
       when my house goes down,
       when she really collapses
       and there's nothing left
       to see
       under the steaming rubble
       and ash settling on stone.
But I got tired of waiting,
I couldn't see anything
anymore anyway.
So I drove to the market
to buy lipstick and mascara.

—*Anne Ziebur*

56

Song

Nancy A. Pietrafesa

hen it was time to leave, I sang to my house. I composed a sort of Gregorian chant as I moved through rooms, stuffing precious items into paper bags. Letters from my grandmother. A christening gown. The glass cylinder of shiny stones my husband and I had collected on a lonely beach long ago. I sang loudly: "Good-bye bedroom, where we comforted one another and where we dreamed. Good-bye little boys' rooms that cradled my babies and held them through the night. Good-bye study, where I hung my heart. Good-bye rooms, where we lived the quiet moments of our lives." My chant lapsed to a lullaby. "Good-bye house that held us safe. Good-bye, sweet house. Good-bye sweet house."

Through the night, tossing in an an unfamiliar bed, I imagined my house fending for itself, like the little houses in the Laura Ingalls Wilder books. At dawn, when the fire had moved south of our neighborhood, we awoke, and my husband dialed the bulky hotel phone next to the bed. In the dark, I heard the clear sound of my voice on our answering machine. After the beep, my husband whispered, "Hello house, we love you." □

from FIRENOTES

To be part of a major natural disaster is a humbling experience. As I live the aftermath of the disaster, I find myself in awe of nature, as another human being woven into the web of life on earth. It is this feeling, I think, that binds us together to support one another and to listen to one another's stories. We were not really different from the medieval peasants in the stable hearing the sagas of their clan over and over again. Or the veterans of a war, reminding each other of their miraculous survival.

—*Margaretta K. Mitchell*

The Family Bible

Linda Morris Fletcher

n the living room, looking out the window at the elm leaves blowing horizontally from the tall trees on my street, I said to my husband, "Now and then God does something right. Imagine if that brush fire had been today."

Instantly I wanted to take back my words, for our house and our street darkened. "What's that, Mom?" a daughter asked. I already knew, even before we got outside to look up to the black sky and watch the orange ribbon lace its way toward us, flaring—boom! and boom!—the way Christmas wrapping shoots up in the fireplace. House after house.

Willy, never an alarmist, thought we should put some things in the car—"Just to be on the safe side." Our three daughters ran into the house that had been built shortly after the 1906 earthquake. It wasn't just the house's old age or the swing on the front porch or the huge trees that spoke safety; it was the five big pumpkins on the porch. I always buy pumpkins too late to enjoy them. We carve them and they rot. This year was different. Those bright orange beauties were on the porch well before Halloween. I was organized. I was in control.

After gathering the photos and my mother's silver, I ran two-stairs-at-a-time up to the third floor to the family Bible. It is a very large, very heavy leather-bound volume that is fastened together by two sets of brass clasps. Its six inches of pages are gold-leafed at the edges. It was given to my great grand-parents, John and Annie Aitken, as a wedding present. Their marriage is the first event recorded in spidery, faded brown ink in the sacrosanct middle pages.

The names of John and Annie's ten children are written on the "Births" page. The older ones were born in "W.T.," Washington Territory. My grandmother, Evelyn Violet, the tenth and last, was born in Washington State.

Her sister's names were Madge and Blanche and Effie Mae and Maude. What happened to Maude at the age of four is recorded on the "Miracles" page: A tree felled by John came down on her just at the moment she tripped into a hole. But she died of diphtheria at the age of ten. My Gram used to say God wanted Maude early—he just missed the first time.

Gram said everything else, too. How not to marry on the rebound like she did and end up with a ne'er-do-well like George, who left her and my mother during the Depression. And how if I didn't brush my teeth, I'd end up with false teeth like hers-and really it was worse losing your teeth than your husband. And about keeping blisters clean because President Wilson's son—with the best possible medical care!—died of blood poisoning from an infected heel blister. And about the Indians who came to her mother's door hungry—so quietly they startled us!—and how everyone wore white on the Fourth of July, and how there was nothing better in the world than a ripe peach.

And about the fires.

The forest fires that threatened the homesteaders, and made such a mark on my Gram as a girl that she told the stories over and over, living with my family as she did all the years of my growing up. How she did go on about those fires—the biggest ever in Whatcom County!—got the barn, but saved the house and the animals. Wet blankets on the roof. Buckets from the stream. How they ran, even as little children. Oh, how hot and tired and frightened, and oh, you'll never know anything like it!

I held my Gram close to my chest as I brought the family Bible out to the porch, past those absurd pumpkins. That couldn't have been me. I knew better than to think, even for a minute, that I could be in control. After all, I had Gram's own shaky handwriting on the "Deaths" page, recording the death of her daughter, my mother, at the age of forty-seven. I had yet, after twenty years, to record my sister's death, at the age of twenty-seven. I just hadn't gotten around to it.

I set the Bible on the backseat of the car. I looked at the orange ribbon working down behind the Claremont. And I looked at my house. Which way was the wind blowing?

My Plan Had Been

Stan Washburn

My plan had been to spend Sunday afternoon building shelves in the auditorium at The College Preparatory School, where I am a trustee and teach drama. Smoke from the fire became visible from our home in Berkeley while I was finishing lunch, but I had the impression that it was on the other side of the Caldecott tunnel, so there seemed no reason to change my plans. As I drove into Oakland the smoke was increasingly impressive. All through Berkeley and Oakland the streets were lined with people looking toward the hills.

When I got a clear view of the hills I saw Hiller Highlands was involved, which was close to the CPS campus. Had I known how close, I would have returned home, like a good citizen, to be out of the way; but being almost there I wanted to reach the school if I could.

CPS is built on a rectangular piece of land in a ravine, with its foot resting on Broadway, Brookside along one long side, its short upper end bounded by Eustice, and Golden Gate forming the other long side.

A block from campus, at Broadway, I encountered a police barricade. I turned up Ocean View to approach from above and began to meet cars filled with people and belongings careening downhill, lights on, swerving in and out on the narrow streets. Towards the top of the ridge above the campus the smoke was thick and low-lying, like fog, whipped along by the wind and creating a deep yellow twilight.

I reached the upper entrance of the school, where a slow-moving police car with a loudspeaker was ordering the immediate evacuation of the neighborhood. The streets were full of hurrying people in cars and afoot. The smoke was dense, and flames were burning on the other side of the campus. I parked and slipped in the side gate. It was about 12:30.

The campus appeared deserted. In the distance I could hear the evacuation announcement, now louder, now fainter, over the considerable noise of the wind in the trees. The only immediate danger appeared to be the slope behind the library at the corner of Eustice and Golden Gate, where the trees and undergrowth were aflame. There I found Tim Bindler, a former fighter of forest fires, who had seen the smoke from his home, come to help, and selected CPS as his battleground.

Using a garden hose, Tim was holding back the fire twenty feet or so above the building. I located a second hose and a spigot to hook up to, and took up station a few yards along the bank from Tim. Between us we could cover most of the immediate front along Golden Gate. It was brisk work, because as flareups would break out here and there, it was almost impossible to force our way through the bushes to get at them, and the range of the nozzleless hoses was short. Up toward Eustice, the bank was too high and our hoses too short for us to accomplish anything. We began to fear that the fire would get past us, moving along the top bank of the campus and down again among the buildings.

At this worrisome moment student reinforcements appeared: Ollie Chaine, class of '91 and his brother Alexis ('93), Zack Jarvis-Wloszek ('95), Ben Witte ('92) and his father, Richard. Ollie and Ben are drama students, Richard is the father of a drama student, and for the rest I can't say a handsomer thing than that they are worthy of being drama students. Dean of Students, Andy Dean, arrived along with a stranger who introduced himself as Ethan. With the added personnel, we were able to extend our front past the library, up to Golden Gate, and straight up the hillside to Eustice. We contained the fire within this corner of the campus, extinguishing smoldering fires at the bases of trees and suppressing new breakouts.

Due to the puny reach of our hoses, we had to move around a great deal. Working in difficult terrain meant each hose operator was forced to nurse the line from place to place with heart-breaking consumption of time. The hoses caught in unburned brush, and we slipped on the muddy slopes. Altering the point of attention more than a few feet often meant bringing the hose back down the hill to the library path, then laying it back up the slope directly below the new danger. Water pressure was low, so each time a hose was moved it went dry until it had filled up and began to flow again.

At some point a new fire opened up between Brookside and the student commons. When we discovered it, we had to disengage a hose from the library and transfer it to this new front. As before, the hose filled slowly. The fire by now was only six or eight feet from the building, licking at the eaves and gutters, which were full of dry leaves. Embers were falling thickly into the combustible detritus along the wall. I backed under the eaves and stamped out embers until the water, in its own sweet time, began to run.

It was one of our many lucky breaks that the ravine created a backdraft, so that even with the strong westerly wind at the level of the treetops, the air at ground level was still or backwinded, which made our fire move slowly.

Things were not going as smoothly beyond the campus. Houses just across the street to the east and south were burning fiercely. Pine trees caught and flared wildly, burned to the trunk in thirty seconds. The smoke was thick, so that visions of the flames suddenly appeared and disappeared. The severe wind fanned the fires, knocked down rubbish from the trees, and lofted sparks and embers long distances. Oakland fire trucks parked in line in the middle of Golden Gate were flanked on both sides by erupting houses and had to effect a hasty retreat.

All around us we heard intermittent thumping explosions as the gas tanks of cars cooked off. Chemical bombers, an assortment of thundering, clattering old propeller planes wheeled through the billows overhead. The thumping of helicopters was incessant. But my impression of the time was of an almost eerie silence.

Falling embers were becoming a serious problem. The campus was littered with eucalyptus leaves which catch fire easily. An ember landed on the shake roof of a house just across the street. It burned slowly in the hip of a dormer for some minutes, the sort of pretty little fire you would cook wieners over. A gallon of water would have put it out, if anyone could have gotten a gallon of water to it. Then it took hold, spread rapidly across the roof, and engulfed the house.

As the campus eucalyptus trees began to catch high up where we could not reach them, I worried about the trees flaring and making instant retreat imperative, or dropping burning branches, against which there was no warning or protection. Tim observed, though, that although the trees were smoldering, they were not flaring. They continued to do so all day and into the night. For unknown reasons they never really burned.

Embers falling on the roofs of the school buildings were out of range of our dinky hoses, but the composition shingles did not catch fire. As the embers fell on us, however, we had to hose down ourselves and each other. We also soaked our shoes so we could walk about through the burned-over areas, which were frequently still burning just under the ash.

At one point I climbed up the embankment to Golden Gate, and then, mistaking my way, started down a route which turned out to be a thin layer of mud over a sheet of coals. It was slippery and there was no turning back. I slithered down through it to the path at the bottom, windmilling with my arms for balance so as not to have to put my hands down. Shortly afterward Andy Dean, maneuvering another slope, fell and broke his ankle.

Once I found myself on the slope, straining for foothold in the slimy mud, clinging to a bush for support and hosing a small flareup, when a tiny black and gray bird appeared. Almost indistinguishable from the mud and burned twigs, it hopped forward and bathed itself in the side-spray from my hose.

For a time we held our own, or actually gained on the fire, but by 2:30 our state changed for the worse. The fire by the library had moved around us and was now burning across the top of the campus, involving the trees there, and turning down towards the structures. Both sides of Eustice were burning and nearing the line of school vans parked along the curb. Suddenly Andy, whose ankle was by now badly swollen, came and said the Fire Department had ordered us to evacuate.

Evacuation was premature from our perspective, but in the face of a direct order, there seemed no choice. We called in everyone from the firelines, and hastened from the smoldering slopes to the center of the campus. It was infuriating. I had pictured our evacuation as a mad dash through the campus with the buildings on both sides hopelessly aflame, not this prudent withdrawal on somebody's say-so.

Andy passed out keys to the vans. None of the students had driven in, and Andy's truck, parked on Brookside, had been destroyed, so there were plenty of drivers. They ran up to Eustice and drove the vans—hot to the touch—out of the flaming street to a rendezvous we thought would be safe.

I returned to my car and watched the vans go by. With the confusion, and my not knowing everybody, I was not certain who had actually been present. Because it seemed to me there had been more people than I had seen leave, I went back down the side stairs into the campus and made a careful search. There was no one to be found. Now the smoke was thicker, and the light seemed

thin and flat. The wind wailed in the trees as the fire gained hold. The buildings were still untouched, but it was clear that without resistance they would soon be in flames.

Andy and a couple of students, looking like naughty school boys, came hurrying down the steps; they had stashed the vans and sneaked back through the police lines. Whoever had ordered the evacuation was not prepared to enforce it. Soon almost everyone was back. We set to work again, in considerably worse shape than before: the fire was well established along the slopes facing Eustice, and houses at Eustice and Brookside were beginning to burn fiercely. The street here is narrow and overarched by trees, offering a quick bridge from the burning buildings to the school.

At this moment Oakland Fire Department Lieutenant Bob Morgan appeared, followed by several of his firefighters from Station #3. They brought with them a three-inch hose, which they attached to our campus hydrant and snaked up to the upper campus. Oh, after those piddling garden hoses this was grand! It blasted out a magnificent column of water which bent back the brush, struck up sheets of mud from the slopes, showered dead leaves in all directions, and utterly smothered the firefront which had threatened to overwhelm us. It reached the trees and the roofs, and over the roofs to extinguish smoldering stumps and logs on the Brookside flank of the Commons. After the agonizing frustrations of the garden hoses, it was liberating to feel the solidity of it straining against us and vomiting out great gouts of water.

"I like schools," Morgan said. "Schools are important. What're you going to do with all those kids?"

In twenty minutes we had gone from deep concern to guarded optimism. And all this time the houses round about caught and burned by twos and threes, by whole streets. It was a choice:

Morgan thought he could save the school, and knew he could not save the houses opposite.

By 3:30 we were again out of immediate danger, our second crisis of the day overcome. The firefighters departed, leaving us their big hose. Our students had mastered its use under the firefighters' tutelage, and we reckoned the campus fire was very nearly licked.

Still we had to stay alert. Embers continued to fall, and we had several new fires break out here and there. We initiated frequent patrols around the perimeter just outside the buildings, and gathered fire extinguishers from the classrooms, assembling them in the courtyard where we could snatch them up as needed.

Andy, careering about the campus on a mountain bike, organizing and arranging our efforts, discovered a recurrence of an earlier hot spot beside the gym. I snatched up a hand extinguisher and lumbered the weary way down the hill to deal with it. The gym, under construction, had walls but no roof, and its vast open plywood floor, piled with building materials, was just waiting for a drifting ember. There were no hoses or water sources nearby. After dealing with the hotspot I left the fire extinguisher on the floor, a tiny red dot in the middle of the vast expanse.

One of the houses directly opposite the upper gate on Eustice had caught fire at the eaves in front, over the door. It seemed a slight fire, and slow. We had no action just then, so I went across the street, took up the garden hose in front of the house and played it over the roof. Most of the flames seemed to be extinguished, but in one place the fire had spread into the attic. When I climbed up on the porch railing and tried to stick the hose up into the burned-through part of the roof, it partially collapsed in a shower of burning wood, and I had to jump clear. I gave up the house, one more among many, and returned to campus. Some

of our people later went across the street, knocked a proper hole in the roof, put out the fire and saved the place.

Late in the afternoon water pressure was lost at hydrants on the street, but not at ours on campus. Later I learned that when our campus was built, the school was required to bypass the inadequate local system and hook up to a distant main, at vast expense. This saved the school, and a good slice of the neighborhood downwind.

When their own hydrants ran dry, OFD came to hook a line to ours, disconnecting our three-incher in the process. We couldn't argue with this decision, but we felt considerably more vulnerable without our big hose.

Shortly before sunset three helicopters in line, with waterscoops suspended, passed the school at low altitude, heading for Lake Temescal, to fill up. Great plumes of smoke surrounded them, the sun behind them dim and orange, and everything appeared in a golden tint. Only "The Flight of the Walkyries" was wanting.

By now we were all caked with mud, drenched with water, gray with smoke, lip-chapped, red-eyed, worn out. I had begun to feel all this unwonted exercise. My knees were sore and I felt I had only so many sprints left in me. We needed help, and the task of getting it was complicated by the absence of any communications with the outside: all the phones were dead, and we feared that if we sent out messengers, the police would not let them come back. But through the cordon of authority, materialized Hans Crome (European history teacher). He bore Andy away in his car to a far-off place where telephones still worked and our wants could be made known; and in time bore him back again.

Around five o'clock the organizing hand of Janet Schwartz, Assistant Head, began to manifest itself. Reinforcements, consisting of parents, students and staff who were able to argue, bully, or sneak their way through the police lines, began to arrive to relieve us. We were famished: bottled water, soft drinks, junk food, assorted fruits, every species of sandwich, cookies from Mothers and cake from Cocolat were delivered. Cellular phones, flashlights, extinguishers, spare hoses and shovels were brought in. A kerosene lantern appeared and was suspended from the balcony in front of the math office. It looked like a particularly fortunate sort of hobo encampment. Andy Dean lay on a sleeping bag, his foot bandaged, iced and elevated on a cushion. With his myrmidons in conference about him, he suggested a whimsical parody of the famous painting of Wolfe expiring on the heights of Quebec.

I regret that we had no anthropologist present to study the firefighting costumes adopted by professional people. Unpolished loafers and white gym socks seemed universal. Khaki slacks and neatly ironed oxford button-downs worn without ties, open at the collar, and rolled up to the elbow, were a pretty general response to a desperate situation, but we had one or two lumberjack outfits, complete to the tall lace-up boots. By now everything was de-ironed, and smoke and mud had softened all colors to various muted grays. The effect, rather than the initial outfit, had become our uniform.

I had thought myself very calm through the afternoon, but I noticed that my language had grown uncommonly foul. I have learned from experience that this means I'm excited. Yes, I was excited. It had been great fun, exhilarating in those hours, when there was so much of importance to be accomplished, to be on the spot and able to accomplish it.

When the cellular phones had been cleared of urgent immediate business, I was able to call my wife. She had received a message earlier in the day

that all was well, but it was reassuring to hear one another's voice. The next day, when I saw reruns of TV coverage, I could see why she had been concerned.

We were winding down now. Clint Wilkins, Head of School, had finally managed to get in, finding a bored policeman who wanted to see the sights to give him a ride through the lines. Clint settled onto one of the cellular phones, and produced sheafs of lists. This normal scene brought home to me that we had won, barring really atrocious luck.

The wind had died around sunset and the terrific advance of the fire had stopped, but all around us the shells of houses continued to burn. When the houses themselves were gone, their gas lines burned on with little dancing flames here and there in the night, all over the hillsides and up the streets.

We gathered under the lantern in the courtyard, surrounded by the dark but wonderfully untouched buildings, and ate. We listened to the radio, which had much dither but little firm information. It was reported on good authority that the school was gone, which gave us some amusement. We chatted and admired the full moon, then went out patrolling for flareups or looters.

Around 1:00 I stretched out on the floor in the language office on a thin carpet over concrete. Thinking, This is gruesome, I fell instantly asleep, and awoke at 5:00.

At that hour it was still quite dark, but cooler and almost clear: stars were visible. There was no wind, and few vehicles. It was still, very peaceful. The gas fires had disappeared, and only dwindling heaps of debris still burned and glowed. I ate a turkey sandwich, drank some juice, and felt like a new man, except that my knees were stiff and sore.

The helicopters had discontinued their flights at nightfall, but resumed them at dawn, clattering overhead. We reconnected the great hose and bore it up and down the street, resoaking our slopes and those opposite. Monday would be a better day for fighting fires. The wind was lighter, and from the sea; it was cooler and more humid than the day before.

LETTER FROM MARY HUTTON

On that terrible Sunday, as we waited to see if the wind would shift in our direction, our daughter kept calling us to come over to her house in San Francisco. Our granddaughter called to tell us to be sure to bring the cat when we came over. A friend called to tell me to be sure to take my poetry. My husband asked if I'd remembered to put the fire insurance policy with the things to take. (I hadn't.) And, finally, another friend, who lived closer to the fire than we did, called to say that she and her husband had talked it over and had decided the best thing they could do, since they were both elderly, and he was still recovering from heart surgery, was just to leave. They would not bother about taking anything but themselves and would go north through Richmond and over the bridge to San Rafael for the night. Then, if the fire got their possessions, they would still have each other.

FIRE-PROOF

You are printed on the insides of my eyelids,
waking, or sleeping,
always closest to me.
To turn, and hold you like a child, a lover,
a friend, long married to,
my father, and my mother.

—*Mary Hutton*

Rescue 1000

Emily Jurs Sparks

unday morning the sky started to darken despite its being a hot, blue day. An infrequent Santa Ana wind blew from the northeast, much hotter and drier than usual. All the doors were banging, and already the yard was thickly strewn with debris fallen from trees. Leaves and parts of branches were coming in the open windows, piling up in corners of rooms.

My son Ben had just arrived home from his fall wilderness job. After cleaning out his car, we set out, as usual on Sunday morning, for the big dog park on the bay. From the freeway we could see lots of ominous smoke but continued with our outing. The dog-walk was not fun; soon we could see, far, far away (along the ridge of the hills?) even flames, and the vast, black cloud almost covered the bay, with the water now a gun-metal gray. It began to feel immoral to be out anywhere, trying to have a normal day.

Back home, we discovered the orange glow in the sky getting closer, and the television and radio were broadcasting serious trouble. Fritz and Ben clambered around our roof with hoses, pointing spray also at our huge, drought-compromised redwoods, copper beech and other trees. All the use of our precious water seemed a waste, as the wind was surely evaporating any amount in an instant.

In haste and controlled panic, I gathered clothes, photos from the walls, photo albums, jewelry, papers, early schoolwork and drawings by the children, Tabitha's journals (a whole big basketful), her photos, my stories, and I loaded the cars. I had to catch all my birds in the aviary and put them into a cage that could be transported, knowing the task would be impossible if our electricity were knocked out. The poor dogs could smell smoke, knew something big was awry. Molly lay in my path right on the threshold of the open front door so as not to be forgotten, while Chloe rushed around with me, back and forth, every step of the way.

Some of the families on our street were packing everything they could, some were hardly doing anything. People stood outside in knots, looking at the sky and shaking or nodding their heads. Children played catch and generally tried not to get involved. Everyone repeated the same news reports. Five arms of fire fanned over north Oakland and part of Berkeley, devouring blocks at a time and jumping two freeways. What better firebreak could there be than four-and six-lane freeways?

Friends and relatives were being evacuated every hour. Our friend Skip brought carloads of valuable goods and his parrot here. He left his fancy motorcycle in our driveway, which gave our frightened house a slightly jaunty look. A helicopter over my parents' neighborhood told them by loudspeaker to be ready to evacuate. A message on our answering machine told us that Fritz's dad, Bob, and Dorothy had already lost their house. Their whole hill was the first to go. We later learned that the jacket Bob was wearing had holes burned in its back. They had fled with the dog but not the cat, who turned up two days later at Broadway and 45th, at least two miles away, with four burned paws and singed ears and whiskers.

In the afternoon, after helping Fritz on the roof and watering the neighbor's roof, Ben put on his sturdiest boots, found gloves and a thick old hunting jacket, and went downtown to volunteer. Of course I was not sure when I would see him again, especially if we were forced to leave, but there was no stopping him.

He found the officials turning away hundreds of volunteers. They were not organized for a disaster of this magnitude and did not want anybody to get in the way of the effort. They were afraid of liability, so, incredibly, felt they could not use these people.

After some dickering with the officials, Ben and a group of loosely organized men, most unknown to each other and all unknown to Ben, decided to go anyway. Hopping into pick-up trucks, they drove toward the Oakland-Piedmont border. They agreed to give themselves a name to sound more official, as their appearance was anything but, not having uniforms like so many other groups come from afar. "Rescue 1000," they called themselves and only because of that were able to get through fifteen to twenty police barricades over the course of the night.

It worked, and they later were to hear police radios telling each other that "Rescue 1000" was coming and to let them through. (Of course, there were people volunteering in the neighborhoods, but it was from the downtown staging area that they were, ironically, not allowed to volunteer.)

In Ben's group, almost all but Ben had firefighting experience and were coming from elsewhere. One man gave Ben an extra helmet, and many had brought tools with them. Ben directed them to the

Blair reservoir in Piedmont, from which they walked over Harbord and down Moraga Avenue to the corporation yard. The three city firefighters were so glad to see these volunteers that they could not express their thanks. This is very heavily wooded, steep terrain, too slippery, I guess, for the firemen in their city gear, because they stayed on the flat, while Ben and the "rough and readies" combed the hillsides.

This precipitous canyon has no houses, and the workers spent the night clearing swaths of land, digging and turning earth, handling the hose (the big-nozzled hose requires three men at once), and being rewarded faithfully with drink and sustenance from the Salvation Army. At one point during the middle of the night, they returned to the main staging area downtown and were given, along with the "official" workers, plates of spaghetti, coffee, eggs and sandwiches. After this interlude, out of the original twenty-six of his little group,

Ben and seven others returned to the fire and worked, axing and picking, shoveling and hosing, until after dawn.

All night I stayed up watching TV for news and warnings, privately staying up to welcome Ben home. At 1:30 a.m. I had one of the surprise-thrills of my life: I was changing channels when I came to Channel 4 and saw a small woman interviewing a black giant. I was riveted by what he said—how, from time to time, it felt as though they, the firefighters themselves, were on fire, it was so hot. The giant I had not recognized at first was my own son, black from soot; I only recognized his voice and his talk! He was safe! I had tuned in on the end of his report, so had not heard where he was, or his name, but soon I learned that two of our friends had seen him, too, so I knew I wasn't dreaming.

The next morning before 8:00, I walked the dogs, and met Ben driving up the hill, soot-blackened from head to boot-toe. Instead of being fatigued and drained, he felt "great," high from the satisfaction of his work. When I asked him how close the flames were when they felt so hot, he said, "Two feet."

Today it is hard to imagine ever getting back to normal. Much in the newspapers covers the fire. There are endless lists of addresses and names, pictures one will never erase. I am, as we all are, overdosed and saturated, but in a voracious way, cannot seem to get enough. Because the burned areas are on the hills, and the hills backdrop to much of Oakland and Berkeley, the fire damage is visible from virtually everywhere lower down. So even when it is temporarily out of your mind, it is not out of your view. You have only to look up, or ahead. You cannot avoid it. Three thousand homes destroyed, in the very city where I was born and grew up, cover a lot of space, geographically as well as emotionally.

Lines of Defense

Terry Shames

I

s a child, I lived on the Gulf Coast. Hurricanes were our disasters. My family stayed home during more than one, and I remember being mesmerized by the sight of thick-trunked trees swaying to touch the ground. I left it to the grown-ups to worry about the damage the trees would do if they tore from their roots and hurtled into nearby houses.

I am still struck senseless by the disruptive power of calamity. All morning as the fire approached, I saw my neighbors filling their cars with possessions, while I merely wandered from room to room in my house, not seeing what was there. I couldn't imagine what would have meaning for me out of the context of our home; sterling silver, china, art, electronic gear.

My husband, David, kept plugging away at practical tasks. He backed up computer files, packed the home videos and business papers. He watered the roof, drove the cars to a safe place down the hill so we could get to them if our uphill escape was cut off.

Intermittently observing the progress of the fire from my attic window, I saw a frightening and impossible situation. Fireballs leaped across the hills above the Claremont Hotel, driven by the intensity of the heated wind. The sky behind the hill reflected eerie light, as smoke blocked the sun with a screen that in only the most rudimentary way reminded me of fog. David and I speculated whether the loud popping sounds every few minutes were trees exploding into flames, or houses bursting open, or the gas tanks of cars blowing up. Toy helicopters dropped thimbles of water, and airplanes spread fairy dust fire retardant, only to have the fire lick it up and thrive on it.

Our four-year-old son, Geoffrey, had not thought to question why he was being allowed to watch one video after another. Eventually, saturated with this unaccountable treat, he emerged from the cocoon I had created for him in our bedroom, to realize that something extraordinary was happening. Why else would the sky be so dark, when it still felt like day? When I explained the situation truthfully and, I thought, calmly, he responded without equivocation: "Let's get out of here." He was echoing the friends who kept phoning, asking, "Are you okay? Why aren't you gone?"

I forced myself to pack photo albums and a few overnight items, but I was unable to actually leave. I was waiting for the fire to make the decision so eminent that I would magically know what I should take with me.

At 4:00 in the afternoon, when the wind appeared to be shifting back to the west, toward our house, and the fire department suggested we evacuate, I said, "No, it won't reach here. Let's have hot fudge sundaes instead." So my husband and my son and I sat down and ate greedily.

I was glad when police came at 6:00 and urged us to leave. Despite the fact that David opted to stay, to discourage looters, I wanted out of the smoke, away from the gawkers who had gathered by the score on our street, away from the hum of my underlying, unacknowledged fear, from the sight of such destruction. But I was kidding myself that it could all be left behind.

II

I was so jumpy when I arrived at Carol's house in north Berkeley with Geoffrey and our dog in tow that I couldn't sit to watch the TV coverage of the fire for more than five minutes before I would be back at the telephone, trying to reach friends, gather still more information. Relieved by good news, stupified by bad, I struggled not only to maintain calm, but to impose order the best way I could. Finding that my son's preschool would be closed the next day, I organized a play group at a park so that he and his friends and we shaken parents would have a place to gather. It gave me something to do, a focus in a day that had been beyond focus.

After midnight, saturated by television images, emotions, and the food and wine we had consumed to shore up our flagging spirits, Carol and I lugged a mattress upstairs from the basement for me to sleep on. Later I awoke imagining I was choking with smoke, my throat raw. I padded to the open window, and believed for a moment that I would see the fire's perpetual dawn lighting the sky to the east. Then I remembered I was far from the light and smoke.

III

How could I not have known that I need to protect the people I care about, way past the bounds of the possible? I give advice, try to fix things for my friends, hover over them. The night of the devastation I could not comprehend how, since I had been spared, I had let this terrible thing happen to them. I fantasized that I had stopped the fire, prevented it, by a psychic vision. That I called the fire department and told them their fire was not quite out from the day before. I told them, urgently, to send a crew up to the mountain, to snuff out the fire. I saved my friends' houses. I saved the house where Barbara held my baby shower. I saved Margarite and Nicholas' house, where I had stood in the yard the day before, admiring the cat. I saved the tree the cat was playing on, and the cat. I saved the houses of members of my mother's group and book club. I saved the art collection of the woman who remodeled my house. And the homes of my two friends who were away. I saved Linda's photos and the sweaters she knitted before her hands were crippled with arthritis. I even saved her husband's business files.

IV

The morning after the fire I marked time in the park with my son and his friends. I felt raw, like someone with new skin. The adults talked in dazed, tentative voices. One man playing with his son on the swing seemed quite giddy, carefree. He told me he was stalling, prolonging the time before he found out if his house was still standing. The children were as always hungry, thirsty, full of energy, whining, giggling, mischievous.

Finally my son asked, "Mom, when can we go home?" and I heard it resonate all over town, children asking, "Mom, when can we go home?" to parents who had already explained that there was no more home, at least not the one they remembered. How could I have told my son there was no more Big Bear, Peter Pan bedspread, books, bulletin board, wolf poster, dinosaur lamp, stuffed animals crowding the bed, green and white rug, baseball cards, bow and arrow, broken swords?

"Let's go," I said. When I had spoken to my husband this morning before he left for work, he had told me I would have trouble getting around the barricades to return home. But maybe we could walk around them.

In the van we kept circling, homing in, and were finally allowed to park two blocks away. I later discovered that when we walked away from the van I forgot to lock it, leaving my most precious possessions vulnerable, as if now that they had escaped the fire they no longer mattered, or that they had been rendered invulnerable. I was to carry them in my car for over a week, unable to transfer them back from a mobile to an immobile environment.

On streets usually lively with traffic, an eerie quiet reigned. Smoke still drifted skyward from the ruined hillside above. Geoffrey and I struck off for home. The houses still standing all around were like an imaginary landscape. How dared they be so bold?

SHELTER

In the junior high gymnasium,
we have everything we need.
Free phone calls, hot food,
counselors to monitor
our feelings. Outside,
there's the orange sky,
fire behind the Claremont,
and in nightmare sleep,
the hurry, hurry
that pushed our bodies
this morning. Here,
in the room for families,
children call for their pets.
Some people weep,
but Marianne's bird
sings on her pillow.

—*Sonia Saxon*

III

AFTER THE GREAT FIRE

We return and sift through
the ashes of our homes
archaeologists
how many centuries
have passed
since yesterday

—*Ellen Cooney*

LETTER FROM GABI'S MOTHER

Like many others, I lost my home on October 20, 1991, in the Oakland Hills firestorm. My 18-year-old daughter, M. Gabriela Reed, her boyfriend and I were at home when the fire broke out for a second time that morning, only a block from our house at the end of a quiet cul-de-sac. At the last minute, I remembered the dog and ran back to get him. In that brief second, my daughter left in her car ahead of me and became lost in the smoke and ash. Her boyfriend and I escaped; her body was found on Charing Cross near the police officer and the others who were killed on that narrow, winding road. In one short hour, my life, as I knew and understood it, was changed forever. On the day of her Memorial Service, I wrote the enclosed verse. For me, this was very significant, as you need to understand, I've never written verse or poetry and don't even know how. This came from some place within me that is hidden and remote, and that I can touch only under such extreme circumstances.

GABI

They did not make me view the body, rather—
sit quietly between the two grief counselors.
"Yes, this is the ring I gave her
for her sixteenth birthday."
"This is her class ring;
this is the ring her boyfriend gave her."

All that remains of my beautiful, brown eyed baby
is in a small gold box
amongst the flowers on the table.
I touch it gently
hoping to feel some part of her.

I think her soul floated away with the balloons,
but the pain remains, growing.
I want someone to tell me why this is
and how to make my life normal again.

—Susan G. Spoelma

Sermon

Stewardship Sunday, October 27, 1991
Rev. Samuel J. Lindamood
Piedmont Community Church
Text: "Peace I give unto you; my peace I give unto you.
Not as the world gives, give I unto you." John 14:27

 his sermon is not as I had planned it. It will certainly be one of the most existential sermons I have ever preached. It will also be quite emotional, so please bear with me. I will keep my handkerchief handy here and you can keep yours handy if you like. I'm sure we will get through it o.k.

I had planned, as some of you know, to get up in the pulpit and say, "After thirty years this is my last stewardship sermon. Thank God!" Can't do that now. I was going to recall for you the best of those thirty stewardship sermons and let it all hang out for one final time. Can't do that either. They're all gone, burned, with all the rest of my sermons. Maybe it's just as well. They are already coming back to haunt me.

The grandson of one of our good members said to his grandmother, "You remember, Grandma, Sam preached that sermon on the Kingdom of Stuff. He said stuff isn't important. You don't really need all that stuff." That's true! You really don't need all that stuff.

This is a ball! A little blue ball. I know you can't see it very well; actually, it's a little blue squash ball. It's a ball that you can throw back and forth. Throw it very quickly in our house and it will be returned just as quickly in the mouth of a beautiful little brown Pomeranian. That little Pomeranian, Kodiak, and this ball are all we have left.

I used to think that I just loved the Pomeranian; now I adore the Pomeranian. I am one who does not like the use of the word "bonding" very much. Bonding has become far too trendy for me, but I suspect that I will now be bonded to that dog for as long as he and I live, because we came through the fire together.

I don't tell you this to make you feel sorry for me. I really tell it to you for the lessons that can be learned by all of us. And I think I probably ought to tell you enough details of what happened so that three hundred people won't come up to me afterwards and each one ask me what happened.

Ann and I were in Arizona for a little R and R and came home early Saturday night about seven. Our daughter told us there had been a fire across the canyon, but it had been controlled, and everything looked fine as far as our eyes could see. Ann and I got up early, as always, and since this was a Sunday morning I didn't have to preach, we took a walk around Hiller Highlands. It was, at that time, a beautiful day. The wind was just beginning to blow, a light zephyr; beautiful sky, lovely day. Little did we know what lay ahead.

It was our oldest daughter's birthday, so we were going to have a birthday party that night and I was going to buy some salmon and other groceries. Hopped in my truck but the battery was dead, which turned out to be a blessing. I took Ann's car instead and went to the store. When I returned, it was about 10:30 a.m.

My good neighbor, Tony, asked if he could help me jump start my truck. I said, "Sure." So I got out the cables and we tried. Nothing happened. Wouldn't work. Battery was absolutely dead. But because of that effort, we were standing outside. Soon we were joined by eight to ten other neighbors, all watching the smoke in the distance. The wind was now blowing something awful—hot, swirling, nasty wind.

We stood watching the smoke in the distance, seemingly a long way off; actually, it was just over in the next canyon. We watched and watched, none of us ever dreaming that it could or would soon be upon us. As we watched, in came the chemical bomber, quickly followed by at least four helicopters. Surely they would control it.

As we watched, the sky just above the crest of the hill got pinker and pinker. The smoke was unbelievable. The pink changed to red and then the flames crested the hill. I have never seen anything like it. It struck two homes with shake roofs about the distance of a quarter of a mile away. Let me tell you, folks, don't ever have a shake roof. Those houses exploded. It was like they went, "Boom."

Ann was down inside the house, totally unaware of what was approaching. I ran to the upstairs door and literally screamed at her, "Get the dog! We're leaving!" She didn't believe me; she thought I had panicked. I backed the car out of the garage and left the motor running. Now there was thick, swirling, heavy black smoke and hot ash in the air. I saw an apparition through the smoke. A voice spoke to me and said, "Reverend Lindamood, is that you?" I said, "Mamie, is that you?" (Mamie works for us twice a month) "What are you doing here?"

"I am on my way to catch the bus. I have been down at the neighbors."

I said, "Mamie, get in the back seat of the car

quickly." She did.

I screamed again at my wife to come. She had been talking on the phone to our daughter, still not believing me. But this time she came, bringing the dog. The whole incident took no more than seven minutes but now the fire had "topped out." It had blown sparks over the top of us and started a fire coming up the hill on the other side. Everywhere we looked there was fire. It was a conflagration within a matter of minutes.

We drove down the hill in only one car. Thank God the battery was dead on the other. Some people took two cars and got separated. One made it, one didn't. We made it, together with our dog and Mamie, driving down the hill through flames on all sides. Hollywood itself could not have produced anything wilder than what we saw in those few minutes.

We are deeply grateful. We are grateful to be alive. We're grateful to you. The outpouring of love and sympathy we've had from all over the country has been overpowering. We only wish that everyone who got burned out had as many friends and had as good a church community as we have had. The support has been wonderful. We are grateful.

I want to say to you that in the next few weeks and months there are going to be all kinds of analyzes of that fire and what happened. How could it have been changed? I don't know how the fire started. At this point I don't care how it started. The last thing I care to do at this point is blame someone.

As far as I am concerned they could have had a regiment of United States Marines out there with hoses and they could not have stopped that fire. It was so fast and furious as to be absolutely incredible. No one, no fireman, no policeman could have stopped what happened. Those guys and gals gave everything they had. The policeman who died, died within a block of us. In fact, more than twenty of those who died, died within a block of our home. Most of them never had a chance. They had no warning. I was outside and saw it coming. We are grateful to be alive.

It does make you think of what's important to you. Ball, dog. My wife left her diamond rings, everything. That's all we got. That's o.k. I wouldn't trade Kodiak and his ball for any of those things. A lot of funny stories begin to surface about people who had more time to decide on what to take, anywhere from ten minutes to a half an hour. Obviously we had no time, but lots of people did.

Most of us have played games asking ourselves what we would take with us under such circumstances. It was a game played many times during the days of "values clarification." But let me tell you, when the fire is right there and the smoke is whirling in your face and choking you, you do the damnedest things (pardon my language).

People take things that they wouldn't ordinarily take; sometimes it's totally irrational. My wife got the dog and grabbed the ball, and I'm glad.

One of my best friends had a while to get out. He went into his house, got his banjo, grabbed his golf clubs, fixed a martini and went out to the car. That was more staged than irrational, but when the disastrous moment arrives, you simply cannot believe it. You're in a state of denial, looking out and thinking it can't happen; it can't come over here. And then, "pow," it's right there on top of you. We are grateful to be out.

About fifteen years ago, two paleontologists, one from Harvard and one from the Museum of Natural History, came out with a very interesting theory. They hypothesized that most species down through the eons of time change very little. But, once in a while, in a small isolated population, there will occur a major change in a very short period of time. They called this "punctuated equilibria."

That is my favorite idea for stewardship. That same phrase is applicable to our lives also. Most of us just go along doing the same things for a long period of time. We call it being in a rut, but we could call it a state of equilibrium. Then, something happens, tragic or otherwise, and our lives are suddenly changed. We make quantum strides in a very short period of time. We go from people who only worry about ourselves and what we have, to people who are concerned about others and can reach out and give love and support.

There is often such a gap between what we say and what we do. There is an enormous difference between our professions of commitment and what we actually give. In our state of equilibrium we sit on our pocketboooks and wallets. When something punctuates our equilibrium, we can make great strides and find ourselves able to give far beyond what we had imagined. It doesn't happen often enough, but it does happen. Too

often we wait till illness, death or some form of tragedy strikes.

This is an awfully long introduction to what is going to be three short points. I thought and thought about what I wanted to say. Under these circumstances it comes down to some pretty basic stuff. There are three words, used greatly in the Old and New Testaments, that represent the gospel for me. Nothing new and exciting but much more important than stuff: love and joy and peace.

I. LOVE

Love is probably that thing, feeling, gift, happening, that can pull us out of our equilibrium faster than anything else. It is hard to describe what the expression of love means to people. Unfortunately my experience as a pastor tells me how hungry people are for love because they don't get much of it.

You and I celebrate knowing the love of God in Jesus Christ. We call it God's grace. But we experience the love of God mostly in each other. The capacity to reach out and touch and hug can change lives, in spite of all that has been said recently about sexual harassment.

We went up on Friday afternoon for the first time to survey our home; it was beginning to rain. There were a bunch of neighbors there, staring at what looked like nuclear holocaust. As we looked and ached and cried a bit, people that I hardly even knew came over and hugged me. Interesting, isn't it? All of us live in neighborhoods of some kind. We have good neighbors and those that we think aren't quite so good. A few of them are prickly, right? Take the guy on our street who's a little prickly. Who cares? Hug time, folks.

When it gets right down to it, there is nothing that expresses the caring of human beings for each other more than just hugging someone. That expression of love between human beings makes all the difference. At a time of hurt and pain, nothing else matters, really. It's a pain to have to go through what we have to go through, but what really matters is that I've got my four daughters, I've got my wife, I've got my dog and I've got this damn ball.

Guilt is not a good motivator for giving. All sorts of groups in our society mail us things and make appeals to us. Often they use pictures and other methods to make us feel guilty so we will give. Unfortunately that kind of guilt doesn't last very long and we give no more than we have to to relieve our guilt.

What really produces a difference in giving, a punctuation in our equilibria, is the desire to reach out beyond ourselves. The realization that love is what life is all about changes us in little and in large ways. When we suddenly have impressed upon us that we easily could have lost life and loved ones, it changes everything. We are open to the possibility of giving in a fresh, new way.

II. JOY

I had all of my sermons scheduled until Christmas. My Thanksgiving sermon was to be entitled, "It's Good To Be Alive." I don't have to preach it now, 'cause I'm telling you today, "It's good to be alive." A young student four doors down from us died. I feel terrible about that. She was apparently asleep until the last moment and never had a chance; we didn't even know she was there. We were lucky. It's good to be alive.

Lost all my poetry. I've taken years, writing four books of poetry, all in longhand, one for each daughter. All gone. All gone. I can scarcely believe it. But I still remember well e.e. cummings' great poem that begins, "I thank you God for most this amazing day." And remember the psalmist who said, "This is the day the Lord has made. Let us rejoice and be glad in it." It's good to be alive. There's a lot of joy in that.

Joy is not some kind of feeling that has to do with your facial expression. It has nothing to do with smiling or being nice. It has nothing to do with going to parties and having fun. Joy is a feeling deep down inside you that wells up and makes you glad to be alive and thankful for a new perspective on life and the assurance that God is really there.

It's good to have another chance; there's joy in that. We celebrate that every Sunday morning when we make our confession and then hear the assurance of pardon. Then with joy we can face another day, another week, another year. There are thousands of people in our town and city who are going to need some joy in their lives. Let us help them to believe in the possibility of a new day and a new beginning.

III. PEACE

Finally, there is peace. Peace is a beautiful word. The old Hebrew word, "Shalom," is a wonderful word. We don't have anything that quite matches it. Often times we cry, "Peace, peace," and there is no peace. But peace is not simply the cessation of hostilities. Peace isn't withdrawing somewhere and having a moment when things seem quiet and peaceful. Like joy, peace has to do with what's inside you. Peace is that very special calm you have inside when all hell is breaking loose outside. Peace is the gift of God that enables you to believe things will be all right when the evidence seems to indicate the opposite.

I have to tell you we have spent hours with an insurance adjuster. Really nice man. A "good ole boy" from the South. His is a hard job. Goes all over the country where tragedy strikes. His stories go on and on. Nice man. Tells us we have to list everything we have ever owned—when you bought it, what you paid for it, what it's worth today. Who cares, you say to yourself. Well, ultimately you do.

It's a terrible process. Sit down with your four daughters and a computer and start going over the house, room by room, item by item. It's a terrible process. More than a little frustrating. After a while it is easy to get depressed, I don't care who you are. You can only talk about it so long and then you get depressed. You have to get away from it.

It is in the midst of such that the word peace takes on new significance. In the midst of the painful memories, in the midst of despair and depression, peace is not a word that rolls easily off the tongue. I would like to have simply said, "Peace," to many people, but it wouldn't work. It wouldn't work for me, either. Peace is a gift from God that has to come from inside out, not outside in.

Paul talks about "the peace of God that passes all understanding." That phrase rings true but I can't tell you why. There is no way that I can rationally explain that kind of peace at this moment. I can only tell you that I believe and feel that that peace is available for all of us in the midst of whatever may come. One of the greatest prayers ever written says, "Grant, O God, that in prosperity we may not forget thee, nor in adversity think ourselves forgotten of thee." Perhaps His greatest gift to us is peace in the midst of adversity.

Let me conclude with a couple of stories. The other day a wonderful lady walked into my office. She's here this morning and I don't want to embarrass her. She came in with tears and left in tears and left me in tears, but wonderful tears. She said, "Sam, there are moments like this when I always think of Hymn 421. I said, "I know the hymnbook pretty well, but I'm sorry, you'll have to refresh my memory. She said, "Hymn 421 is, 'God

will Take Care of You,' " and with that she placed an envelope in my hand and walked out. It was a very generous gift. But her greatest gift was the gift of peace that comes with Number 421.

The rest of my life, wherever I go, I will always remember 421. Reminds me of that awful old joke about the guys in prison. They had been together so long that they just called out a number and everybody laughed. A new guy came in and called out a number, but no one laughed. He didn't understand. They said, "You don't know how to tell a joke." I may not be able to tell a joke either, but I will always remember Number 421.

My daughter had a very interesting dream the other night. She dreamed that someone ran up on the hill and spread Miracle Grow, and it all grew back just like it was. Her dream is a wonderful metaphor, but Miracle Grow has to do with love and joy and peace. Miracle Grow has to do with you and me reaching out to people who are in despair at this moment and sharing with them like you have shared with me and my family. Miracle Grow has to do with loving each other in a way that will enable our world to rebound and rebuild. Miracle Grow has to do with those wonderful old words—Love and Joy and Peace.

Jesus said, "Not as the world gives, give I unto you." Believe it!

To All My East Bay Neighbors

Karen Klaber

The Monthly

am one of the thousands who lost my home and precious possessions in the firestorm. Since *The Monthly* is delivered to every home in the neighborhoods of Oakland and Berkeley devastated by the fire, I have felt a special, almost familial connection to you. Now I reach out to you in deepest sympathy and empathy.

I understand that part of the grieving process involves telling the story of your loss. I lived at the end of Gwin Road, off Broadway Terrace, near Skyline. I had a panoramic view of the ridge where the fire began. On Saturday, I witnessed the relatively small first fire on the ridge. Late Sunday morning, I was horrified to see a ferocious blaze twenty times its size spreading so quickly. I knew immediately the lives of my best friend, her husband and their son (my godson), who lived on the ridge, were in jeopardy.

Through frantic phone contact, we planned our escape and decided to meet at Peet's Coffee, near the Claremont. I fled without thinking of my own possessions. We all managed to escape with our lives, if not our homes. My heart goes out to those who were injured and especially to those who lost loved ones. Thank God for the many heroes and heroines, without whom many more homes and lives would have been lost.

I woke up this morning (Tuesday, October 22, two days after the fire) in an unfamiliar room and cried my eyes out. Though I am eternally grateful none of my friends died in the fire, it finally hit me that I would never again feel the comfort and pleasure of my home, family photos, art collection (many of them my own creation), personal journals and poetry. Some possessions—my piano, jewelry, clothes, books and furniture—can be replaced, but many of my dearest personal possessions remain forever irreplaceable.

I didn't own my home, but it was a spectacular, one-of-a-kind place. For the last six years, I had painstakingly created an atmosphere that supported me emotionally. My home was a refuge to me. It was also an important expression of myself. When mine was destroyed in the fire, I felt like part of me died.

I have always enjoyed living in the East Bay. We are a community of intelligent, broad-minded, creative people. Our salvation for the future will come from our inner strength. We will rise again. For now, let's feel the pain, tell our stories and accept the love of the generous friends who have rushed to our sides with offers of food, shelter, clothing, money and shoulders to cry on. My staff and I send you love and consolation.

Dear Mom And Dad

Jack Compere

ow it's like living next to a grand ocean liner that has sunk. The feeling from the neighborhoods burned is that huge and ghastly. The undersea comparison prevails elsewhere—driving around our saved neighborhood, shopping at the saved market, all of it done slowly and carefully and mutely as if at the bottom of the ocean.

Extra care is taken because the memory of the chaos is strong and it's easy to make dumb mistakes because your mind is elsewhere, and you're careful in public because you know that many of the people you see have recently been sifting through ashes, and you don't want to do or say the wrong thing.

Today has been a day of reclaiming our sweet little home; cleaning the floors and dishes and clothes and so on—at least that's what it looks like we're doing. Actually we're just touching and cherishing the home we nearly lost. Scattering lots of birdseed out on the deck to make up for the time we were evacuated.

We don't need convincing that the fire was on its way to our place. We're still finding "firebrands," the burning detritus the fire throws out ahead of its path, which look now like chunks of coal but are practically weightless. I was sitting with Carey on the hill opposite our house on Sunday night as we watched our hill catch fire. The stand of pines over our neighborhood was backlit by the flames. The heavy clouds of smoke, moving eastward, were reflecting deep reds, now and then joined by a spiral of black smoke that signaled a house burning. We watched motorcycle cops darting here and there on our streets, looking for people to evacuate. Their white headlights were the only lights, because power to the houses had been lost long before—in fact, we couldn't actually see our house because the fire put our side of the hill in deep eclipse.

I believed that the next thing we'd see was the fire catching the canopy of the trees on our hill, so we left. As Carey and I drove away from this sight, I leaned out the passenger window and said goodbye to our house.

When we'd gotten up Sunday morning, I told Marvel, "It's a bad day for a fire." The winds were hot, strong enough to push the pines around, and needles and bits of branch were falling heavily. Because we're from Southern California, it was easy enough to recognize the Santa Ana conditions, and when I took the Sunday paper out to the deck, I also brought the binoculars, to watch for fire.

Marvel went down to visit Mindi. On her way out she met our next-door neighbor Andrea, and they briefly talked about the frightening conditions. Andrea was leaning on her car as they talked, and at one point she said, "The car is breathing!" The car was rhythmically pushing against her because its bumper was up against the pine next to their garage, and the pine was swaying enough to rock the car.

I read the paper (which included a small article I've saved, covering a grass fire that was "extinguished" on Saturday—the origin of the Sunday fire). Marvel called to say that there was a fire up by the Caldecott Tunnel. I turned on the TV, but there was no coverage; I got out the portable radio and took it back to the deck. KCBS said that a grass fire was sending up a lot of smoke, and houses in the Grizzly Peak area were threatened.

I was still on the deck, writing, when Marvel and Mindi came in. "You should see the smoke!" Marvel literally turned me around (I had been watching the hills you can see from the deck) and pointed. The sky to the north was filled with rapidly billowing cumulous clouds.

Marvel and Mindi took the rabbits down to Mindi and Carey's house. I kept the radio on. I watered the roof and everything else. Carey arrived and we began to put stuff in his truck and my car. I kept on watering. Carey raked the roof of the house and the garage. A couple of times in that afternoon I told Carey to go on ahead—to let Marvel and Mindi know I'm on my way. Each time Carey would fix me with a stare and say, "Jack, I'm not leaving 'til you leave."

Now and then, through the afternoon, he and I struck out into the neighborhood towards the smoke to see how close the fire was. On one of these forays we went all the way to Virgo, and for the first time got a good look at the fire, which now seemed to be everywhere. Houses exploding, the roaring noise; trees would first catch flashes of fire throughout their limbs, then settle down to burn seriously. The fire walked up the stairs of a house at the bottom of the hill we stood on. Behind us, a lady said her house was going to go. On the way back to our house we told many people that the fire was close, and it didn't look like it was slowing down. We untangled a hose and tossed it to a man on his roof.

I got several phone calls. (The phones worked off and on throughout.) Some were friends wondering how we were doing, and, as the afternoon progressed, I got increasingly urgent calls from Marvel. After having seen the fire, I felt silly with my little garden hose. The 10% humidity and winds made everything dusty and dry seconds after being sprayed. The chemical bombers were coming in over our house now, to dump on the valley behind our hill. The police were broadcasting commands from helicopters and cars to evacuate the hills. At 5:10 Marvel put it most succinctly: "You might save the house but lose our relationship."

Just before we left, one of our neighbors said that if everyone stood on their roofs and watered down their houses, the fire wouldn't be getting as far as it was. What actually stopped the fire was the cessation of the hot winds, suddenly, late at night, as if someone flipped off a switch.

We visited a friend of Carey's in Piedmont whose house was also being threatened. This group of folks thought that anesthesia was called for. I thought it was weird to get twisted behind alcohol as the flames blasted towards your house. Carey and I again went down the street towards the fire, and this was as close as I got. A helicopter dumped one of those 400 gallon buckets of water within 100 yards of where we were—it appeared out of the smoke, dumped the bucket, and disappeared again. It seemed silent, but that was only because its engine was nowhere near as loud as the fire.

We saw Carey's friend much later that night, sitting all gloomy on Carey and Mindi's front porch; we asked if his house were still there, and he said he didn't know. Then he called back to Carey, asking for any alcohol Carey might have in the house; this time he needed anesthesia because he'd hurt his ankle. Apparently, while on his roof, he yelled down for someone to throw him a beer—he caught the bottle but fell off the roof.

That night we stayed at a hotel because Carey and Mindi's place was full of people. When I finally closed my eyes I saw burning trees, burning houses.

Monday morning we got a call from Mindi, who said someone had seen our house still standing. We cried, and I didn't believe it. In my mind's eye, it had burned down many times by then. All of Montclair was closed, so Marvel and I took a back route around the hills, until we could get a peek at our hill. There sat our house, the banner waving slightly over the deck. We stared and stared.

Later in the day, Carey and I took his truck the back way and parked it as close as possible, then we hiked up to our house. It was an eerie trip, because no one was back yet. People had left faucets turned on, so the silent walk was punctuated with the sound of running water. People had left odd things in their yards. A file lay in the middle of a lawn.

School pictures—two long class pictures, black and white and obviously old, were sitting at a curb.

As I write this, I haven't yet gotten the nerve to take a walking tour, although I'm curious to see if the house where we helped the man on his roof still stands.

Yesterday Marvel took flowers to the Montclair Fire Department station, to thank them.

Today, I filled five bags with pine needles and branches and so forth; as I worked I realized that I was moving faster and faster, as if only by working could I stay ahead of disaster.

Love,

OAKLAND AFTER FIRE

Tonight we're climbing the backside
of these familiar hills
hunched against the slant
of moonbeams. At the crest
we find the fireline that finally held.
Below, the slopes fall away
alien and smoky, as if the moon
had gutted the landscape.
Nothing stands
but chimneys stretching their shadows
across the soft
slow breathing of ash

and black bones of eucalyptus
not even moonshine
makes fair.

      —Taylor Graham

Fire Journal

Tobie Helene Shapiro

October 31, 1991

t is Halloween. I always looked forward to the incongruity of it all; tellers at the bank gigged up like jesters, hobos or ghouls. On the street, cars go by with four small fairy princesses in the passenger seats, and a clown driving. On the sidewalk, men and women in business suits walk together, discussing transactions with a six-foot-two rabbit carrying a briefcase. People dress as they want to be, or as they aren't, as an unexpressed part of themselves.

I am a dentist, but today, Daddy is a big baby in diapers, and Mommy, the realtor, is a pirate.

When it gets dark, clusters of little ghosts and huge boxes with legs, escorted by their parents, troop up the streets, stopping at houses festooned with pumpkins. The door opens and a bright rectangle of yellow light glows out into the night as the candy flies into the paper bags.

All this, tonight, is after the fire. It was a sort of uncostuming affair, in which we shed all our clothing, possessions and footprints. The hills shed their trees, and what remains are chimneys and foundations—a wall or two, caved in on top of rubble. The hillside is dark now as we drive by on our way somewhere else, trying to remember not to go home. And when we do go home to try our luck at miracles, each time, it is like being awakened from a nap—the house is still not there again. Or maybe it is like being put down for a nap—next time, it will be there. The neighbors all return, just like we do. Now we are all visible through our hedges and trees, our walls and our pulled drapes. It was all a dream.

The tour busses are coming through, as are the families from San Jose who have packed a lunch and a camera, to watch us sift through our ashes. I weep over the remains of a stamp pad, and I grieve over all the sorrow, the anguish that passed in that big, beautiful, awful house.

November 8, 1991

In any tragedy, there are lessons to learn. That mitigates the pure horror: a kind mouth around the teeth that bit you. It gives the whole thing a more complicated expression and smoothes a place for it to settle in your life—an odd feeling of belonging. Also it differentiates tragedy from travesty. Nothing is learned from travesty—or even worse, the wrong lessons: learning to lie next time; not to feel next time, not to learn next time.

One could learn to die a thousand small deaths of the soul, cremate the vision while saving the eyeballs, on the way from here to there, from A to B, from "a glimmer in your parents' eyes" to an obituary next to the weather report.

This fire is full of lessons: about ending, about continuing, about the grace of surrendering, the kindness in goodbye, about beginning, about choosing to create from nothing over recreating from the ruin. I have learned that I am not materialistic. I suspected it, but it is nice to really know. Hell is not poverty, it is purposelessness.

No. I take it back. Poverty IS Hell. They are both Hell. There is a lot of Hell. There is hell of Hell there, dude. And all of it, every chunk of defoliated foundation, every layer of ash, every mystery motor, every fleeting fragment of undocumented memory, every curled window slumping into itself, every erased footprint, is a lesson. Thank God for tragedy. Thank God for tragedy. Yes. I was blind when I came here, but now, I can determine light and shadow! Praised be the Lesson!

November 9, 1991

When a book burns, its soul is left behind: the imprint of the ghost—the pages black, the print silvery white. Photographs, imprinted once as illustrations, now flicker on and off the page, between the angle of the eye and the source of light, like conversations or small errands that refuse to fix themselves in our memory.

Why did I come into this room?

In this fire, the metal twisted out of shape. Some melted into pools. Hollow things collapsed in upon themselves; glass exploded, melted, bent; most everything just entirely evaporated—the table and chairs that became a shadow, the armoire and silk robes I may have inhaled this morning. And the books: the paper burned, but the books retained their souls. They are uninjured messages now, as hard to read as a life.

Nothing in the ruin is wet. I think the firefighters made no attempt to douse our private flames. We pull pottery out of the ashes, newly fired and usually with barnacles of glass or plastic. The books crumble in our hands.

In the basement, I had a case of 1954 dry Sauterne, bought cheap in 1975, now worth maybe over a hundred a bottle. As near as I can figure the miracle, the bottles exploded and leaked their dear contents onto a box full of my creative writing:

letters, poems, song lyrics, maybe even an early short story or two. The liquid soaked the pages, and when the fire moved through, it was all too drunk to burn. (Yes, if it hadn't been for that pint of whiskey I drank that night, I would've stiffened up when I fell down those stairs, and I wouldn't be here telling this story to you.) And so the only written work that survived with its body intact is an eight-inch-thick pile of my creative product.

I had forgotten about it, really. Had said goodbye years ago. And here it is, stinking from the fire, my own handwriting staring up at me, the ink leaking into the page, from the midst of the black and ash under the burned-out harp of the grand piano that fell through the floor on top of it all.

"An excellent use of that wine," said Xena, a woman of tremendous internal proportion. I pried the whole soggy mess up out of the negative mold it had made in the hill of ashes and mud, and carted it off to the edge of the excavation, by the cracked and blackened back stairs and retaining wall. It is our fireplace mantle. On it, we display our new heirlooms: the pedals from the Ivers & Pond upright; silt-filled and misshapen silver baby cups from Aunt Anne and Uncle Kuo; charred and brittle metal wall hangings—a fish and a flying pig, their paint burned off (the fish was a gift that replaced the pig, that was a gift to my husband David. It had been destroyed almost a year ago when my father, in a zealot's frenzy to dispose of a room full of ribbon and wrapping paper, somehow took it from the safety of the mantle and tossed it into the fireplace in the house that Jack built); metal tools without their wooden handles; cloisonne, bubbled and pock marked, unrecognizable—all looking as they would if they had been drowned 500 years ago when the ship sank to the briny bottom.

November 13, 1991

This is what it is like to have a home. On the morning of the fire, I was home by myself. I sat at the kitchen table with my feet on another chair. And Garrison Keillor was being very silly on the radio. I had a cup of coffee in my hands and luxury on my mind: "I will take a nap." This is part of my Reasonable Goals program.

I am not a reasonable woman. David had taken the twins to the park. For all three of them, leaving home was going to the big silver car to stalk the trains and the Merry-Go-Round. They left, saying, "Mama, you rest," and they just never came back.

For Ben, who had stayed overnight at a friend's party in Lafayette, it was a negotiation about when he would have to come home, and the pain-in-the-ass confusion in the morning from his fanatic father and stepmother about where-were-you-and-why-didn't-you-call.

Alex had left home for U.C. Santa Cruz only a few weeks before. He had come up once to clear out more of his things. He'd said goodbye, just didn't know what it meant.

I was home alone. And my feet were up. There was a table. There was a chair. There was my Great-grandfather's desk, at which he sat studying Torah or delegating business matters to my Great-grandma Rose—or maybe he bought that desk and moved it in as one of the first rudimentary fixtures of a new life in a new house after the quake and fire destroyed the old house on Folsom in the city built on hills. There was a microwave, a hexagonal wall of windows looking out upon the blue, unsinged sky. And behind me, I heard the wind chimes making their heralding bell song, in an unseasonably hot and temperamental wind. There in the kitchen and in the rest of the house, all our things, our things, and things of things, from scotch tape to dictionaries, artwork to notes in the

121

archives, sweaters and letters and sheet music and dirty clothes—these things were all there, too.

And the really nice part of this being what a home is like, what a house is like before a fire, is that they were all, including myself, defined by function and aesthetics. This heats things; that covers the floor with something soft; this is a tool I use to open a door, and this makes my eyes and soul happy just to look upon it; I make sounds with this; this presses an image onto a page if pressed first onto an inked stamp pad; Feyna's head lies upon this pillow at night; David and his big boys slept in this tent under the stars, far from here—a cloth and aluminum home folded in the basement of our home; and I am the mother and stepmother, the poetry fountain (temporarily capped), the shopper and the chef, the cellist and the art maker, the weeping and laughing wife, the fierce advocate for my children, the nag, the family jester and family Jew, holding a cup that holds the coffee that the

woman drinks because a human has a variety of thirsts.

And after the fire, all of this must be defined, not by function or aesthetics, but by Loss, by Price. "How much for this rubber stamp that says, 'Deceased'?"

"Oh that! I used it to mark the envelopes of junk mail to be returned to rude senders and have my name removed from their mailing lists. It gave me much pleasure."

"How much is it worth?"

"It was an expression of my humor and disdain for the invasion of mass-produced bad taste and unethical behavior."

"How much for a rubber stamp?"

"It helped define my aesthetics and purpose."

"Could you replace it for $5.00?"

"Not like that one. No. No other like it. The C had a nick in it. Never came out quite right. It was great!"

"O.K., $3.50."

How much for this desk? How much for this writing tablet filled with black ink and treadmarks? For this stair, for this hankie, for this piano, for this woman on the chair with the feet up holding a cup of coffee because a human has a variety of thirsts?

I do not produce much Money.

I am not a large loss.

November 17, 1992

When you take out insurance, the insurance can cover the tangible loss—the paper and the ink, the paint and the canvas, the husband's and the wife's salary, but not the ingenuity nor the future. The shell of life can be paid for, but it will contain nothing. The wife was pregnant when she was called away by an unsportsmanlike act of God.

We are sorry for your loss. This check will cover the clothes she was wearing, her wallet, her last hairdo, even the makeup, and two year's salary, plus a Nanny, a maid, and a cook, if that was your division of labor, until you can find a new one. But if you want her pregnant, you'll have to fuck off.

Can we sue for the gene pool?

The coverage is useful, but, damn it, that paper was pregnant. That canvas was in labor.

It is not the things. It is not even the beautiful things. It is not the fireplace; it is the hearth. I lost the time of day, the angle of the sun, the grief between the floorboards, the joke in the kitchen, the argument in the bedroom and the kiss afterwards, the daily, careless faith in the passage of our time and passions as any reasonable man or woman knows it to be.

This is what I said goodbye to on the day of the firestorm, before I closed and locked the back door behind me. No one else had to utter that word. But I did.

I said, as I looked around the house. I took a breath, and I said, "Goodbye."

December 1, 1991

I imagine sitting at a desk. This is a dream, and I can look at myself. In this case, I am in a dark room, leaning over my writing, the white page, shining like a mirror under the moon. I am looking up at myself from the side, almost from behind. It all seems so normal, but my hair is on fire. It is not consumed. It is just in flames. And the glowing page remains empty, even though my hand moves the pen across the page. I write, but no document is recorded.

This is how I am living my life now: tilted away from me, busily knitting ink onto bright pages, the close room, dark around me. I am just doing what everyone does—except that my hair is on fire. I am the light in the room, and I am showing no signs of pain. But nevertheless, this can only go on for the moment. If I turn around, or notice the danger, I'll certainly burn—have to run from the room, igniting my clothes, the curtains.

It is not the real world; the real world is my hair on fire. It is not the real world; the real world is sitting peacefully at the desk, writing.

The reality is that I have no choice as to which is real. It will change with the unseasonably hot wind, blowing from the northeast across the eucalyptus groves and down through the big safe houses nestled in the hill.

THE WEEKEND AFTER THE FIRE

New mums form a brave triangle of yellow
along a street with no neighborhood.
A burnt tricycle waits on a path
to a sill
with no door
to a house
with no walls
under a beam
with no roof
onto a patio
with chairs and a view—
a view with no room—
vertical with chimneys instead of trees.
The wind moves through leafless poles
and flaps Tibetan prayer flags.
Somewhere a door bangs.

—*Judith Stronach*

BARE HILLS

From the train windows
the hills
look like bare hills
You wouldn't know unless you knew

    If my parents had known how to love
    I wouldn't have been in the back seats
    of cars at 15, she said, I wouldn't
    have needed to learn everything
    from the outside in
    Do you remember how the moon
    would look afterwards, she said
    like your own reflection
    uninhabited

Fire, he said
What we might call a resonating
sentence since it offers
more than one voice
landscape
more than one direction
mood
This endless multiplying of the possible
what we might call
beyond control

The narrative's always such a relief
isn't it, she said
Picking up the threads almost without
thinking, or fingers darting
back and forth all afternoon
weaving our histories
When I decided to leave
my second husband
he made a sound like fabric tearing
but so quick I've never been certain

Trains and vehicles moving
along the freeway
People inside waiting
From one moment to the next
you're no longer who you were
No wonder we lose
track, she said

    They were driving up the coast
    pines spearing the slow light
    He said, Why can't you let go
    of the grief behind your eyes
    when you say the word
    father
    She said it
    He didn't turn

to look but they could both hear
grief filling the space
between them like a moon
She reached out to him
in simple need   as if for
a transfer

These faces on the train
maps without ledgers
regions where only
solitary travel can occur

    Most mornings he has trouble waking up
    Something about his life
    But no use talking about it
    He feels stuck enough as it is
    Suffocated, sometimes
    The moment his hand depresses
    the latch on his car door handle
    wild hope floods him
    Almost a kind of fuel
    Like knowing he's not alone in this.

During the chaos of the fire
people had to abandon their cars
flee   flames at their ankles
Fire and music seem mutually irreconcilable

states of being, fire consuming
everything as it does
even silence
But when you look at the hills
from the train
aware that they are bare from fire
your mind supplies a sound
not unlike music
Rare instruments went up in flames
the newspaper reported
Notes in a glissando

    Riding backwards in a train
    seeing what others have already seen
    through windows too worn or dirty
    to sustain illusions
    you can hear bare hills
    fill with grief
    a stillness like the moon

As soon as there's mention
of distance, we ask, *How far? What for?*
As if geography were the same
for everyone   as if maps
could be spread out on the floor
of our desire

She does not desire
to marry again
Two marriages   two failures
enough for any man or woman
to sustain
Nevertheless this urge to join
the body's emptiness
no longer lunar
no longer unrelieved
I don't want to relive
my life, she said
I don't want to lose track of the narrative
by going backwards

These faces on the train home
nothing like petals on a wet black bough
even with the most extreme imagining
It was so warm that morning
not everyone had shoes on
It seemed the world ended there
where the sky blew black
its edge still burning
People lost everything
but what they managed
to fling in their cars

If I stay too long
in one place, she said
I won't be able to imagine moving
I'll feel the moon pull
my blood still   the interior
will have its own landscape
uninhabited

If you got out with your life
that's all you should ask
a woman whose husband died
told a reporter
I could feel her grief on my hands
like newsprint
marking my shame in having claimed terror
that did not belong to me
that I could only imagine
like the interior
behind those eyes
gazing past me
at the bare hills

—*Lynne Knight*

129

Three Analogies

Rick Talcott

1. It's a lot like losing your wallet only 150,000 times worse. Honestly, there is a certain amount of pain in not having the physical things that are connected to memories, but that is dwarfed by the pain of having to replace what's necessary simply to survive in our society. It is possible to live with three pairs of underwear and no washing machine, but I wouldn't recommend it.

2. I lived in Japan in 1969 and 1970. I was constantly learning that I could (indeed, had to) peel off layers of myself in order to get by in a foreign culture. As the time progressed, I discarded what I habitually ate, wore, read, studied, drank, talked about, thought about, and usually did in almost any given circumstance. Each time one of those layers had to be discarded, there was a tremendous fear of will I survive this? I was afraid that the central core of my identity would be lost if I let go of this or that layer. Each time it was a surprise to discover that I was still me.

3. Any man who has been divorced has had practice for being burned out. It really doesn't make any difference whether it's a fire or a soon to be ex-spouse giving you fifteen minutes to be down the driveway with whatever can be carried. □

Survivor

Susan Ito

hen Helen rounded the last corner and saw the white house standing there, covered with a thin blanket of ash, her knees unlocked beneath her. She sat down hard in the middle of the street, biting her lip and staring. There it was: not charred, not wrecked, not blown away by flame. It looked as it had when she had left it Sunday afternoon, when she had staggered to her car with photo albums, clothes, a suitcase full of embroidered tablecloths. All that night, she had sat in a hotel room near the Bay, watching the fire eat way street after street of her neighborhood. It's gone, she told herself. There's nothing left, and now I'm free.

She called her office and left a message on her supervisor's answering machine. "I'm homeless," she said woodenly. "I'm not sure when I'll be back, but I'll check in later in the week." Helen watched the aerial view of the fire jumping from roof to roof. The television was muted, and without the exploding, panicky sounds, it looked like a vibrant, beautiful dance. She had turned the volume down to make the phone call, but after hanging up she realized she preferred it that way, just watching the colors, the men in their yellow firefighting suits, the flushed, exhausted faces of the news reporters.

For forty-eight hours she lay in front of the television, dozing at intervals, waking up to Wheel of Fortune or a soap opera, and frantically switching until she found some cable channel that was covering the fire. She called for room service, ordering all the things that were prohibited on her diet. All the months of reconstruction ahead of me, she thought. I'll be so overwhelmed I won't be able to eat a thing.

Young people in beige uniforms brought trays of potato skins, fudge sundaes, cheese platters and wine into her room. She would be sitting on the bed, stroking the pages of the photo albums, little snapshots of her children in their sandbox in the back yard. "I've lost my home," she said. "Thank God I managed to save these." Some of them put their trays down and hugged her, murmuring sadly and patting her shoulder. She tipped them in twenty dollar bills.

When she heard the neighborhood barricades were being opened on Wednesday, she went down to the front desk and handed her key to the receptionist. "I'm just going out for a couple of hours," she said. "I hear they may be letting people in to see the sites." The skin around her mouth pulled back grimly.

"Oh no, Mrs. Marks." The woman at the front desk seemed on the verge of tears. "Do you have someone to go with you? You're not going up there all alone, are you?"

"I prefer it like this, dear." Helen said. She smiled. "But I appreciate your concern. Maybe I'll have a nice big order of that fried zucchini when I get back. And a carafe of wine. I'll need it." Then she laughed and went out to the parking lot, jingling her car keys.

The house was standing in the place where she had expected a blackened, burned pit of cinders. From the images on television, she had braced herself for the lonely fireplace, unshielded and naked in the debris. For a moment she stared at the roof of the house, expecting it to burst into flame. Stunned, she pulled her wallet out of her purse. She checked the address against the brass numbers on the front door, just to be sure. Then, without going inside, she got back into the car and drove quickly to the hotel. There she put on her nightgown and had two carafes of white wine.

They turned the gas and water back on three days after that, and people were allowed to return to their homes. Helen told the hotel staff, "A friend has offered to let me stay in her guest room. I'll miss you all. You've been so kind." She left large tips for everyone she had met during the week. She loaded her suitcases into the trunk, laying the photo albums on top. Then she got into the driver's seat and rolled up the window. When she was on the freeway, driving towards the scarred blackness of the hills, she screamed, a long, terrified

animal noise.

The house was a disaster. For a moment she thought to call the police to report a robbery. But then she remembered that it had been she who had pulled the bureau drawers onto the floor and flung her clothes in all directions. It was she who had emptied the closets and shelves. She looked at the wreck of papers on her desk, and remembered the way she had laughed to herself, saying, I won't have to pay any of those bills, they'll all be up in smoke...

Helen looked out over the pile of still-dirty dishes through her kitchen window. Past the trees in the front yard, she could see the valley below, the chimneys sticking up in the black, empty field. She washed the dishes, humming as the water slid between her fingers. When she was through, she put on her sweater and walking shoes and headed down the hill.

She walked about two hundred yards before reaching the first burned lot with its skeletal remains. The mailbox, standing untouched, had hand-painted letters on the side: PEPPERS. She didn't recognize the name, but it make her think of a children's book from long ago, the *Five Little Peppers*. She imagined that five small children had lived in this house, and were now clinging to their mother in an emergency shelter, perhaps in the elementary school cafeteria. It was an unbearable thought, and Helen crumpled into tears, leaning on the mailbox and sobbing loudly.

After a minute Helen became aware of a shadow that fell close to her feet. She looked up, stricken and embarrassed, to see another woman of about her same age, fifty, standing next to her. She had straight-cropped dark hair with little patches of gray at each ear. The woman was wearing a long brown skirt and wide flat sandals with heavy knee socks. Her eyes were squinting at Helen, and looked as if they might be a little watery.

Helen sighed, feeling helpless, as if the lie had swelled like a river after a heavy rain, and that she was being carried by its force, and that all she could do was look up at the sky as she floated further and further away.

They exchanged phone numbers, and soon they were talking every day, sometimes meeting in the ravaged neighborhood to walk quietly through the smoke-sharp air. Helen made a point of avoiding the Peppers' house after that, telling June, "I just can't bear to look at it anymore." They went on wild shopping sprees, buying clothes and putting the receipts in envelopes on which they had written, GIVE TO INSURANCE CO. It was true that Helen needed all new clothes; the wardrobe that she had used before the fire was two sizes too small now.

Helen drove to meetings of the Phoenix Association, where she stood near the back and ate powdered doughnuts, covering her face with a napkin as she chewed. She looked around at the faces of the neighbors she had never spoken with, the families she had driven past a hundred times without noticing. They were wonderful people. As she leaned against the refreshment table, she listened to their stories and was frequently moved to tears.

She began participating more in the meetings, and volunteered to head the Hospitality Committee. For every meeting, she prepared elaborate trays of baked goods. Homemade eclairs, cupcakes with little candy faces, four different kinds of brownies. She donated a large coffee urn to the Association, and knew what each person liked to drink. She would meet them at the door and then hurry to prepare a cup of decaf, or a chamomile tea. She came to know them all. The lawyer from Broadway Terrace, whom she had seen several times at the video store but never greeted, was now hugging her exuberantly at every meeting. His name was Phil, and he had two young children.

She brought crayons for them to use during the meetings, and they began calling her Auntie Helen.

The lie had been completely absorbed into her body by then. She believed that she had lost her home, that the white house where she lived was being loaned to her by a friend on sabbatical in Europe. She wandered the halls of the house, touching the walls, examining the framed paintings as if they belonged to a stranger. She flipped through the photo albums, curious and detached, looking at frozen images of a family that was not her own. The framed photo of her husband, dead for ten years, grew into the likeness of the professor on sabbatical.

In the spring, construction was begun on June's house. Helen went with her to visit the site, and they took baguettes and wedges of cheese to share with the workers. They sat under the fresh, open beams, trailing their fingers in the piles of sawdust and wood shavings. All around them, orange fields of California poppies lay like handfuls of sprinkled paprika. Each time a floor was completed, June popped a bottle of champagne, and put the receipt from the liquor store in the INSURANCE CO. envelope.

June asked, "When do you think you'll start construction?"

Helen sighed. "My insurance company is run by a bunch of idiots, June. You were lucky. I'll be lucky to get a permit by '94." She put the champagne glass up to her nose and felt the little bubbles bursting in the air. They tickled. "But I'm happy where I am. My friend in Europe. He fell in love with an Italian woman. I don't know if he's ever coming back." She laughed.

They were at a Phoenix meeting, and Helen was chosen to be featured in the monthly newspaper that was being published for fire victims. The room trembled with applause when her name was announced. Her face warm, she stood up and waved at the smiling people who filled the room. "I'm so honored," she said, her voice sliding into a higher octave. "I can't tell you what this means to me."

At that moment, Phil took the microphone. "Helen, your presence here has been an inspiration for all of us," he said. "But you've been so involved with taking care of other people, I don't think many of us are aware of what your loss was. Maybe you could take a few minutes to tell the group your own personal story, where your house was, what your plans are?"

His words echoed into silence as everyone turned to look at her. Their faces were compassionate, expectant. June patted her arm and gave Helen a little nudge.

Helen stared at the microphone in her hand. Its heavy metallic handle,. the wire mesh and thick black cord, made it feel like a dangerous weapon.

For a second she thought of bludgeoning herself on the head and falling into unconsciousness. She lifted it to her mouth, and a horrible, high-pitched shriek filled the auditorium.

Phil gently tugged on her elbow. "Not so close to your face, Helen," he whispered. "Feedback."

She lifted the microphone again and tried to speak. But then the shrieking became real, her own human voice, and she was crying like some sort of terrified tropical bird. "I'm sorry!" she screamed, and her voice bounced back at her, slamming against her ears. "I'm SORRY! SORRY! SORRY!"

June drove her home, holding Helen's hand the whole way. "Honey, if I'd known it would be so painful for you to talk about it, Phil never would've asked you to speak. It was just . . .," her voice faltered, and the car weaved a little on the road. "It's just that we wanted to acknowledge you. You're so special."

Helen didn't say anything. She stared blankly out the window, and the same words kept repeating in her head. I'm sorry. When they reached the white house, she opened the door quickly, almost before the car came to a stop. She leaned in towards June and gave a kiss into the air. "Thanks for being such a good friend."

She went into the house without turning on the lights. The dark air felt comforting after the bright fluorescence of the auditorium. She walked slowly down the hallway, staying in the center of the narrow carpet as if it were a gangplank leading out over a churning, angry ocean. In the kitchen, she found the stove controls and twisted them on, rummaging in a box for chamomile tea. She jumped a little as the circle of blue flames shot up in the darkness. It was the only light in the room. Helen didn't put the kettle down right away but reached her hand out toward the flickering gas flame. The wreath of fire shimmered under her palm.

from UPPER BROADWAY TERRACE

There are no fences
separating families,
no walls protecting
a wife's privacy
a child's fantasies.

I see Mrs. Finch
whom I passed
on morning walks.
Her eyes are blue
and filled with tears.
I never looked this closely
before.

—*Lori Russell*

from Kindled Memories

Deborah Simpson

he cop apologized.

"Sorry I can't give you more time, but we have our orders," he called from the road below.

I turned toward the police van to acknowledge him and then quickly lowered my head to conceal the private trail of sorrow which washed slowly down my cheeks. "Silly," I chided myself, "they were just things, and things can be replaced." I struggled to believe my stoic words.

I began the climb down the hill to the van. It was slippery in places. The fire had destroyed ivy, bushes and branches and left the earth too soft and slick for sure footing. Reaching for a scorched anchor bolt to support me, I glimpsed bright colors. In the ash and dust lay a small fragment of something white, colored with streaks of red and bits of deep blue. I bent down, reaching for what appeared to be a piece of bone china. Ash and dust slipped through my fingers, leaving the fragment resting alone in my hand.

I recognized instantly what remained of an antique Imari bowl my mother had given me more than twenty years before. The design had been scorched in places, but the colors were still as vivid as the memory of the day I had pulled the ribbon off the gift box. My mother and I had celebrated new beginnings that day. I was leaving a job grown tedious and moving on to something new and better.

This day, as I stood among the ashes, I thought of my mother. I remembered the warmth and the laughter of so many years passed; I saw myself in the kitchen where I grew up and called the morning I opened the package and found the special gift inside. Tears rushed forth, but this time releasing joy. I was holding a tangible link to my past.

The horn from the police van sounded one more time. I had kept them waiting too long. I slipped the Imari memory into my pocket and thought of new beginnings.

With love and thanks to Mickey and Dave, Je Neal and Cal, and Connie and Don who have transformed a time of sorrow into a time of hope and new beginnings.

WALKING IN THE OAKLAND HILLS

I never knew the elegance of chimneys,
how they are rooted
deep in the earth,
held by the music
of gravity and quartz,
how they curve upward and outward
swelling around square mouths
blackened by small flames,
then taper, soar,
thrust their weight
against sky.

Women returning to Atlanta
after the war
learned
the shapes of home chimneys.
They never forgot
sooty columns
ranked row upon row
along the ridge
like bloodless soldiers,
answers to questions.
Survival itself was an arrogance then
and each brick
as tangible

as a year without a child.
There are no ashes, no charred beams.
I climb an empty slope, just soil
granular and glistening.
Chimneys stand at intervals
along dead streets.
I come today to breathe
this bitter air, to see
what I have been spared
by changing winds
and minutes.
Someone's concrete steps remain
and a garden statue
made of marble.
I see fire-grey hills
to the north and the east,
knobby shoulders of rock
reduced to essence. Simple geology
except for the push of chimneys.
Chimneys speak of intentions,
the web of names and boundaries
we believed would keep us safe.

—*Sharon Fain*

To All Our Friends

David Kessler

o my left is a stubby plastic box, filled with computer discs. It was one of the very few items I grabbed and ran down the hill with to our Mazda in the late morning of October 20, 1991. It is the only thing in this room that was in our possession before that date. I am working at a computer given to us by our friend Chris Tiedemann, and I will print the result thanks to the donation of her brother Peter. Elaine Zelnik gave us the computer table that holds all these things, and, courtesy of Terry Dean, I am seated while I work. It's cold, this day after the Thanksgiving holiday, but bundled up in the gorgeous sweater that didn't quite fit Michael Rosenblum, I'm snug as a bug. (You start noticing which of your friends are conveniently sized.)

I start out naming names to give you an idea of what our lives are made of now. The raw material is the love of our friends. Everywhere we turn we see the smiles and caresses of hundreds of dear friends, and what may look like a sofa or a mattress at first glance, turns out to be a giant hug in a disguised form.

Three days after the fire, we went up the hill. As far as you could see or walk up the ridgelines, there was incredible total devastation. I wept for the lives and homes of my friends and neighbors, and for so much beauty lost. Beauty of the homes, and the beauty of a whole way of life, of community that had grown over many years. A destruction so complete that no paper or wood survived, where virtually all artifact was gone, reduced to memory.

Walking into this smoking, utterly incinerated ruin, I could not fend off the continuing insight that we inflict a parallel destruction on other societies on purpose, and that our beautiful hills in October were reflections of Baghdad in January. I want to share with all my friends the awful desolation of the sweet piece of earth we inhabited, because I know that until I witnessed this destruction, my understanding of the warfare our society conducts was largely an intellectual one. This, on the other hand, I could smell, see and touch, and I shudder to think of how miserable we would be today if we did not have the love and wealth of a rich society to draw upon to restore our lives and souls. What must it be like in Iraq? . . . Having seen this inferno, I cannot wish any neighborhood anywhere in the world to have to undergo such tragedy, certainly not because of American action.

Nancy and I want to thank all our friends for the love and care we have received. No gift, not even a kind word in passing, was so small that it was not of the most immense value to us.

In peace,

148

Our First-Ever Christmas Newsletter

Frederick Mitchell

 Since the Great October Fire of 1991 was the first class-A disaster of our lives, you'll have to bear with us if we chatter on about it, with lots of *italics*.

Gretta was in the East when it started. The Lord of the Manor had demeaned himself with some late-Sunday-morning vacuuming; it was a damned hot day, with this wrong-way wind, so perhaps shorts should be worn. But while working in the dining room he noticed *orange* sunshine reflecting on the pale yellow walls and carpet. Since fire does not burn downhill, there was clearly enough time to jump in the car and drive up to see what was happening.

The smoke was doing the impossible: boiling up from behind a high bluff at Hiller Highlands and filling half the sky overhead. The smoke clouds were distinctly multicolored—white, brown, and gray, separate but overlapping—and moving faster than anything that big could possibly move. They were being blown high overhead in a rush southward towards Oakland proper. An east-west freeway and the BART tracks lay between, perhaps half a mile away.

After I got home I could see out the back windows that the smoke was beginning to blow *flat*, not puffy, down across the foothills towards the freeway and Oakland. Now it was time to think, "Poor Oakland," time to put the vacuum cleaner away and not change into shorts. Put on heavy pants. Begin to notice many sirens and many helicopters.

A friend, Scott Parker, drove by at about 11:30 with his pickup truck and two friends, wondering if we had anything we'd like to move. "Move?" One of us said, "Gretta's negatives!" and we took the contents of four filing cabinets from the basement up to the truck and he drove them off to his garage. Around noon I decided to move our two cars west to the next block. I didn't hear much from the neighbors; some of us were outside with garden hoses but there was very little said. After a while there were no more cars on our dead-end block.

Also around noon, phone calls began to come in from concerned friends in San Francisco. Steve Kuhn said that the smoke was *very* visible from there, 15 miles west. Vail Maes was worried. The wind was hot and steady out of the east, now seeming to come up toward us from the field behind the back of the house. Those intermediate foothills up behind us were showing some small fires. By now, some part of me knew that Real Trouble was on its way. At the end of the block a police loudspeaker told everyone to evacuate, but the police kept moving.

I had been joined by Allen Leggett, an energetic neighbor, and we got out our extension ladders. Strangers and other neighbors, including a very slender and mournful-looking Syrian woman student, were out watering their hedges. I joined with a burly off-duty fireman who'd been on his way to visit his father's grave when he saw the smoke. We helped our elderly next-door neighbor out of her house and into a car.

At about 12:30 a tiny fire suddenly sprang up steady as a gas jet, out of the bone-dry thin wood shingles on *our roof*, a tranquil little campfire up there, growing out of a large flaming branch that had blown in without our seeing it. I stood at the top of the 28-foot extension ladder. Allen shouted directions to me on how to reach up and point the hose, since I could not see up around the wooden (and pineneedle-filled) rain gutter. No problem! We got that sucker!

Slowly our street was filling with young volunteers. Several were helping at neighbors' houses. I could not even imagine getting garden hoses up onto our roof—it was very high, and if I set a foot on the roof, I might damage the old dry shingles or damage myself. Out through the back, I could now see trees in flames a few hundred yards beyond the field behind our house. At this point I had taken nothing else out, but had moved what I thought was Gretta's jewelry to the basement. I could comprehend nothing that I saw.

Just before 1:00, dark brown smoke was beginning to ooze from between the shingles of our roof facing the street. I had no idea of what might be happening to the back of the house, but I could see through the laundry room that trees closer to the house were now in flames. In the next minutes, the oozing smoke on the roof slowly increased, and smoke was visible in the air all around us. The small fence between the neighbor's and our home was suddenly in flames.

About 1:00 we heard the sound of truck exhausts at the end of the street, and two fire engines came grinding up to our corner. Several neighbors urged them to come up our street. A pumper from San Bruno (40 miles away!) came in first, and the next one, from Lawrence Berkeley Lab, attached its hose to the corner hydrant and drove up the street, paying out the hose in big folds behind it. Soon (forever!) the trucks were connected and had pressure.

There were six available firemen and six houses on fire. The first crew came in with a heavy broad spray through the yard next to our house, and they began to beat back the flames which were at the walls of the Morgan's house and ours. The smoke was so thick that the house was intermittently lost to sight from the street, 30 feet away.

Soon a team wearing breathing masks went in through our front door with a hose, and headed up the front stairs. Since there was no quick and safe way up onto the roof, they had to break into the attic through the bedroom ceilings, which allowed the dense smoke and heat to vent down and out in flames. After a few air holes, the flames must have burned up into the roof fairly quickly and opened all sorts of natural vents.

Someone shouted from an upstairs window, "Turn off the goddam power! We'll get electrocuted up here!"

I heard different sounds: the breaking of glass and muffled thumps from inside as axes hit door and walls, the sound of a full charge of water hitting a burning plaster wall—a hollow hissing rumble. I heard the smoke alarm in the upstairs hall and the little alarms on oxygen canisters, which are pitched, like bicycle bells, to sound when they are empty, all jangling away.

I was going in and out of the lower floor. I saw everything, but absorbed nothing. Heavily suited firemen sloshing around in *our* house! I went upstairs in the steam and water and jumped until I grabbed the battery out of the screaming alarm.

I stepped on and over a face-down bookcase to get into our bedroom. The door was smashed in. (Why did they tip the bookcase over so they couldn't open the door, then smash the door in?) Our middle bedroom window was gone—casing, sash and all. Someone told me that at one point the big pillows on our bed had spontaneously burst into flames, and a fireman swung around with his hose and gave them a blast. Suddenly the air was full of burning feathers, feathers everywhere, even in the tree branches, weeks later.

By 1:45 or so the fire was calmed down. Several firemen were now next door, and volunteers were running most of the hoses. Someone down the hill fired a water cannon toward our roof and anything else that seemed threatened. There was a mist of water everywhere, pouring in from all sides. The street was ghostly with fog and smoke. There were kids in jeans and T-shirts, some barefooted. Many people were wearing folded handkerchiefs across their noses but the smoke didn't seem to bother me. I led Old Dog Daphne out of the confusion down to our car and put her in. Somebody had put Old Cat Amanda into a laundry basket out on a low wall. Three people were petting her and trying to adopt her. I got possessive: "*We* can take care of our cat!"

The last scenes . . . I was called into the living room to find a fireman telling me that the fire in one corner of Kate's bedroom was stubbornly caught in the space between the upstairs floor and the downstairs ceiling. Would I mind if he stood on one of our better chairs while he chopped a hole in the ceiling? I gave what I thought was one of my finest How-Long-O-Israel? shrugs of lamentation, and he jumped up on a nice chair and began bashing with his axe through the ceiling into a bed of coals and small flames. They were beautiful in their way, but there was no time to enjoy them. I hadn't much noticed the fellow crouching on the floor beside me with a firehose at the ready; he let loose with a two-and-a-half inch stream of water, and squelched the whole mess in a big boil of steam, rumbles and hisses. Naturally steam, warm water and cinders shot back over all of us at our end of the fully furnished room; it clearly would have been better if I had put down the lid of the small grand piano I was leaning against.

During the long mop-up, every youth seemed to want to shoot water into the attic of the attached garage since the fire had crept over to it, too. That roof was still whole, so the water had to be sent in by way of a small hinged window in the gable end of the roof. I didn't want to see these zealots smash the window with the water stream, so I decided to stand on a ladder beside the window and hold it open for them, leaning as far away from the water-stream as I could. My sense of delicacy went for nothing except an absurdist spectacle, since the water kept knocking the window out of my hand and slamming it closed. Later I found it lying on the ground, intact.

Late in the afternoon the house was full of people, mostly men, carrying everything outdoors into the hot air. It would be hot all night. By nightfall all of the furniture sat outside. Our friends brought every broom and shovel in the neighborhood and began sweeping and shoveling cinders/lath/plaster/insulation/paper/feathers/ water out the windows and down the stairs. All of the downstairs ceilings were dripping water.

I went to the Dahlstroms and called Gretta at about 6:00. Yes, the canopy bed and the dollhouse had survived. In fact, almost everything had survived, soggy or otherwise. Out our neighbors' window I could see small fresh fires across the freeway south of us.

The night stayed warm. The Leggetts put out some beer and sandwiches and a small TV. My God! We sat, saying little, until well after midnight, half expecting looters or snoopers. Then to bed, removing pants still wet and stiff from all the water, shoes still damp.

The next day the women arrived. Betty Weekes had called a mover. Thirty-one years' (and more) worth of stuff was on its way to storage/cleaners/restorers.

We picked Gretta up at the airport Tuesday night

and took her directly to the house. We carried flashlights and walked through every room, the spaces echoing and plinking with drips of water. No tears, no second-guessing.

Since all of this, both Amanda the cat and Daphne the dog have died of advanced years. We have moved to a rental in the 'burbs, hot tub and all.

The last word has to do with family and friends. As news about the fire got out, we began to receive calls from near and far. Some we could answer; on the busiest days we could only log twenty or thirty calls into our "fire notebook." We remember every call. Even the most perfunctory was amazingly valuable to us. We will catch up to you one by one. This is our start.

Happy New Year!

FIRE SEASON

She soaks a washrag under the cold tap,
wipes the scald of fever
tulip red above the paper white
of her son's small forehead

while the year's first storm
speaks its hushed language.
She thinks of the pond ten minutes away,
hidden in the forest, rain-syllables piercing the surface.
Yesterday, she walked its perimeter,
a mud shore grayed by drought,
saw the neon stitching of dragonflies
and a slate blue heron, fishing in shallows.
It paused mid-step, lifted dark kimono wings,
their silk shadows travelling the ruffled water.
She understood then,
that any landscape could wear sudden flame.

Her son closes his eyes,
curls to a position they both remember.
She touches his spine, the fragile links,
the gold hair matted with sweat.
Nothing burns in this room now
except his molten sleep.
Tonight, she dreams the clouds divide
balanced above the trees and spreading
an ash-coat of light.

                              —Joan McMillan

from Santa Ana's Thumbprint

Gregory M. Blais

I t was the Santa Ana that convinced us to go. Without its push, the fire might take hours to climb up and over the ridge on which we stood. Across the canyon, a mile-wide swath of flames had not only crested the ridge, but was already advancing into the homes at the bottom of the intervening ravine. We stood for a long time amidst other fascinated onlookers. All of us were hypnotized, as if staring into a comforting campfire after dark.   Hot, smoke-filled wind gusted into our faces.

We raced back down the street, warning our neighbors as we went. When certain that all were roused to the danger, we returned to our home for the last time. The cat was the final addition to the pitifully small pile of stuff in the back of the truck. I don't know why, but at the last I went through the house locking doors and windows, as if to keep out the flames. It was 2:30 when we joined the exodus escaping southward.

An off-duty fireman working with a garden hose to save a neighbor's house later told us the incendiary needles from an exploding Monterey pine sealed our home's fate. It was afire by 3:15 and was "already fully involved in flame" when the first engine arrived on upper Broadway Terrace. "There was nothing to do," he said, "but watch it burn and try to hold the line fifty yards downwind." Shortly after, when the Santa Ana retreated into the Sierra Nevada, they succeeded. Four houses past ours.

It wasn't until Wednesday that we hiked overland into the fire area to find out what, if anything, was left. We could easily see that we were not alone in our loss; the devastation ranged across the hills. Any of us could stand amidst our own rubble and see duplicates extending outward in all directions. Nothing but chimneys standing as tombstones over three thousand familial gravesites. Block after block; street after street.

We, like every other soul whose house was destroyed, stood amidst the rubble that was once our home and stared down at it, dumbfounded.

Our senses told us that all was gone, but our emotions refused to "see." We seemed incapable of comprehending the destruction. Our eyes scanned the bleached bones of our home, searching for survivors. Anything at all that might have come through intact. Something to identify this particular smoldering heap as our smoldering heap. Before too long, however, it became apparent that we would have to settle for something less. Only God and maybe a few psychologists know why, but we soon found ourselves taking great solace in identifying skeletons.

"Look, here's the microwave."

"Wow, even the barbecue melted!"

"My God, is that all that's left of a 40-inch television?"

"Anybody find the outboard motor yet?"

On that first Wednesday afternoon while we and some friends stood quietly sifting through the bones, pandemonium reigned around us. Fire equipment of every description still clogged the streets and skyways. Gas and electric utility trucks steadily rumbled by, always on the way somewhere, but never stopping, it seemed, to actually do anything. Police cars and motorcycles constantly patrolled to prevent looting. Up the hill, a bulldozer and men with chainsaws worked to down dangerously burned but still standing trees and utility poles. Smoke still trailed up from countless remains.

Suddenly we four looked up, like deer sensing an intruder. Above us at street level we saw, darting from the safety of one driveway to another, like a honeybee in search of the perfect flower, a tiny red, white, and blue postal jeep delivering mail— seemingly to non-existent houses. We couldn't help but laugh at the absurdity of it all, and the laughing made us feel better. We, like our neighbors, are survivors.

PERSISTENCE OF VISION
for Gayle Zanca

That first blind moment looking out
my mind wanted what it knew, not
these trees like heads of singed hair,

this cemetery of chimneys, the cars—
models made of foil and trampled
by a child who a second later focused

his lens upon an anthill and watched
as the tenants writhed or fled.
But already I can't remember

how it looked. The house across the street
is a pile of dishes fused by heat,
skeleton palm guarding the front

door's ghost, the solitary mantlepiece
stranding a ceramic figure twenty feet
above the ground. A squirrel bobs

and glides among the ruins; the image
of what was refuses to return. Instead,
I see the bones of these hills,

their curves beneath the streets
and rubble, the courses of ancient streams,
and grasses bowing before a clean wind.

—*Leslie Tilley*

No Trespassing

Elisabeth Wickett

h, the East Bay Hills Fire, yes, are you going to rebuild?"

That mindless question from the unscathed, the question with the unstated addendum: "Are you going to make everything normal for me again, so I don't have to think about the voracious chasm which opened up before all of us on October 20th last year—the abyss you fell into, but I didn't. I don't want to know about it or be frightened. I want you to rebuild, to make everything okay for me. Oh, and incidentally, for yourself, too. Won't it be nice when you don't have to think about the fire?"

I reply, "I was an uninsured renter."

"Oh, well, then, you've found a new place to live and you're okay."

I smile bravely and turn away, mourning the ignorance of a person who has a home and cannot pause to acknowledge its meaning beyond plaster, wiring, a roof and a mortgage. Who perhaps has not reflected on the equilibrium and rootedness provided by home, history, family, conversation, guests, projects, associations, music, memories. Is home to those unscorched a conglomeration of tiles and furniture, a television set and lamp, a good address, toothpaste tubes, a garage, and the big one: equity?

Eight months after the fire, I visited what was once my home. The belligerent wire fence and concrete slabs, possessed but unoccupied, of Parkwoods met my eyes. Even the ubiquitous poppies avoided Parkwoods; the place was gray. My mind created an owner's dismissal to burnt renters: Tenants are no more use when they catch fire. Bulldoze their cremains! Lock out the weeping ghosts with delusions of human dignity! If they must stand and stare, they may, like the little match girl, at Paine Webber's ownership rights.

Now heed this warning posted by the Oakland Police:

THIS PROPERTY HAS BEEN VACATED AND IS CLOSED TO THE THE (sic) PUBLIC. THE OWNER HAS FILED A WRITTEN REQUEST WITH OAKLAND POLICE DEPARTMENT FOR THE ARREST OF PERSONS FOUND TRESPASSING OR LOITERING HERE. PERSONS FOUND ANYWHERE ON THIS PROPERTY ARE IN VIOLATION OF THE LAW, THEREFORE THEY ARE SUBJECT TO POLICE SEARCH OF THEIR PERSONS OR POSSESSIONS INCIDENT TO ARREST.
California Penal Code §602(n), §602(1). §607(g)

Posting date: 11/1/91

I stared at the sign. Twelve days after the fire, and we renters, were blotted out as non-people.

I remember being deprived of my own ashes. The walls of my home belonged to Parkwoods, so they claimed all within them, as well.

First, we were kept out of Parkwoods with a tall wire fence and guards. Simultaneously, we were given business cards and a brochure, *We Care and We Want to Help* from the Parkwoods Information and Relocation Office, instantly settled in

Emeryville. Then, Oakland police officers drove us around to look at our ruins from the car window, but we were not allowed to touch or set foot. We were warned—for our own good—off the premises because of asbestos and other unspecified hazardous chemicals. Tenants' rights were discussed between Parkwoods' attorneys and tenants' attorneys, and tenants were guaranteed supervised access with protective gear at the same time that management got in bulldozers and started clearing relics/ashes/debris, depending upon your point-of-view. Tenants resorted to a restraining order. Parkwoods resorted to young people in white clothing who went through the ashes and brought out a bit of the obvious to appease us. It didn't. I was vocally ungrateful. Especially after I was told that I had so much. My mother's wedding china had been meticulously packed under my bed; a few pieces, mostly broken, were saved. Yet, the same woman who told me she carefully went through my ashes and found some china, denied the existence of a four-drawer legal filing cabinet.

I took every shard of my dishes, knives, frying pans, scissors. I took them and I wondered useless wonderments.

Now I am locked out permanently from ground I trod daily for seven years—locked out, threatened with arrest for loitering, for looking at the trees. Locked out from ever searching once more for my beautiful cat. Locked out from even planting a tree, a luxuriant, fluffy green tree to honor my beautiful, beautiful cat, trapped not merely in a conflagration, but in a dispute over property in which owners enforce rights and tenants look through a fence.

It is grievous to put your life on the line and be ineffective.

—Bonnie Cox
Oakland Firefighter

Dear Neighbors

Peter Strauss

I am bothered by nagging questions that keep echoing through my mind. Since I did not lose my house in the fire, why can I not feel joy? Why do I find myself so often in tears, or close to tears? Why, when I reach Broadway Terrace, do I feel such overwhelming sorrow? Compared to those who lost their homes, what have I lost?

I have not lost my house, and all that it contains, that is true. But when we bought this house, we bought the whole thing: the building, the lovely trees whose tops we so enjoyed watching sway in the breeze, the eucalyptus and the Monterey pine, the neighborhood. And over the few years that we have lived here, we have come to treasure it all: the neighbors whose names we did not know, those we would pass as we went for our walks in the twilight; and we have regarded as precious the gentleness of the landscape around us, the sunset filtering through the branches, the elderly couple with their elderly dog, the younger couples with their babies in tow, even the dog who snarled and startled us, the fellow who wasn't quite so friendly, the occasional dilapidated garage.

And so it occurs to me now that while we did not lose our houses, we have indeed lost a part of our homes. And that loss is real, and deep, and because, in fact, we still have our houses, it is hard to justify our grief. But it is nonetheless real, and it is very important that we speak of it to one another, that we share our sadness, our pain and our fear, that we share with each other the experience of terror and despair we faced on that horrible Sunday, some of us not knowing until much later that our houses had been spared. Yes, we still have our buildings, our precious things. Yet nothing can undo the genuine trauma we experienced, nothing can restore to us the previously uncontaminated beauty of those very trees which now we must regard as the enemy, a threat to our security and safety.

And we must speak to each other of what it is like to drive each day out through the fire zone, what it is like for us to expect to see, as we always have, homes and trees and shrubs, and instead to have our eyes and spirits assaulted again and again with the remains of the land—the stark, gray, lifeless hillsides no longer verdant, comfortable or safe. We must speak of what it is like for us to be confronted daily with those grim reminders of what almost happened to us, what did happen to us, and what might happen to us.

We are and will continue to be assailed by what has come to be known as "survivor guilt." Why were we spared, when others suffered so horribly? And how can we feel such sadness and sorrow when others have lost their lives, their loved ones, their homes? We have suffered our own losses, and it is appropriate for us to grieve and cry for as long as we have tears, in recognition and validation of what in fact has been irretrievably taken from us.

The gathering which celebrated and honored those firefighters who saved our streets and homes was not just a block party. It was also a wake. And it was a beginning, neighbors coming together to mourn the passing of part of our home, and to celebrate our survival. It was a chance for us to acknowledge our feelings, our humanity, and to express our compassion for each other and for ourselves. People who had barely spoken to one another in the past, who had perhaps never seen each other before the fire, hugged and cried together. We told of where we had been that day, and what we had done, what we had felt. We met in the context of our frailty, our fragility and our gratitude for those who saved us. We have expanded our awareness of what home is, and who and what it includes. As an old saying has it, "As snow melts on the mountain, new grass grows underneath. With every gain there is a loss, with every loss a gain."

Regards,

IV

Two Gardens

by Nancy Pollock,

as told to Jane Staw

n 1972, we began building our dream home on Buckingham Boulevard in the hills behind Hiller Highlands. It was just after the 1970 fire, and I had a great reverence for the land, for the surviving thick forest of pines and madrones around us. I loved the trees so much that my husband, who was an architect, and I designed the house around the trees. The roof off our bedroom and entryway notched around two fifty-year-old pines. And the casement windows in front of the kitchen sink opened to let the woods into the house. After the house was built, I used to sit in the living room, watching the deer in the springtime come to graze with their fawns on the wild grasses in the backyard. And when I began drawing and painting, I would plan my days so that I could catch the play of sunlight against the trees. In the morning, the needles and leaves glistened and flickered, and in the afternoon, the angle of the sun emphasized the contours of the trunks and highlighted the contrast between the roughness of the bark and the cool smoothness of the spring grasses beyond.

We lived in that Buckingham house for 16 years, our two boys growing up with the trees. The whole time, the back remained natural. It was in the front that I gardened. We collected granite boulders from the Sierras and dug them into the landscape to retain the soil. I gathered plants from everybody I knew: sword ferns from my in-laws' Mill Valley property; campanula from a friend's home; wandering Jew, hypericum, snow in summer from my aunt's Inverness garden; spider plants from a neighbor. Out of these, plus nursery-bought impatiens and azaleas, I created a cascading garden of primarily blue, with contrasting spots of pink and red.

Three and a half years ago, when I moved away to my Gravatt Dr. house, single, my two boys grown, although I regretted leaving Buckingham and the intimacy it provided with nature, I was ready to create an altogether different kind of garden. My new home was situated on a steep southerly slope, overlooking Vicente Canyon, with its soaring hawks and grassy inclines. For the first time in my life, I had full sun, and would finally be able to grow a complete palette of colors. But before I could begin creating, I had to destroy. I had to remove the rotted retaining walls, the intrusive ivy, the unwieldy Scotch broom and anise. As I cleared my land, one foot braced on the downslope to keep my balance, I worked day after day to create a clean canvas for myself. While I labored, the artist in me planned. I imagined my hillside a Monet painting, with every inch a bloom, the hues and textures alive and changing with the seasons. In the northeast corner the pinks and reds of sage would contrast with yellow gazania and its gray foliage. By the front door a carpet of primary colors would meander through a loose formation of rocks. Baby tears would flow down the slope.

During the two months I spent digging and hauling, uprooting and contouring, I gradually realized that my ambition didn't fit the natural conditions of the hillside. I wanted to paint an English garden on a 35-degree slope with poorly draining dense clay soil, natural habitat for the

Scotch broom and anise I had originally found there. If I wanted my hillside dotted with spiky penstamen or lacy yarrow, I needed to create a structure to hold them upright. And if I wanted the dense bloom of pansies and primroses, I would have to import more nurturing soil. So I had a truck-load of topsoil dumped at the top of my hill. And once the soil was in place, I hand-picked three tons of moss rock from which I built the skeleton of my garden, digging the rocks into the hillside in undulating curves to hold the dirt.

As I worked, I envisioned a series of terraces, one flowing organically into the next. But I was too impatient to wait until all the rocks were in place, and I began planting as each small area took shape. I started at the top with a screen of silvery germander and a fifteen-gallon jacaranda tree, full of lavender blossoms and feathery leaves. Next, I created an area of gray foliage, with bush salvia, wall flower, trailing geranium and lavenders.

Around the stones stepping down into the garden, I placed bright green scotch moss, camomile and creeping lemon thyme. As I progressed down the hill, I landscaped with bold garnet and delicate pink penstamen. I bordered the downslope with Peruvian lilies, yarrows, gladiolus, Mexican marigolds and Shasta daisies. For contrast in texture, I carpeted another area with yellow and rust gazanias and white African daisies. Off to the side of my house, I cultivated an herb garden. Flanking the front door, I planted a blaze of primary colors with dwarf primroses, violas and Iceland poppies.

By December I had finished. Each time I stepped out the front door or drove into the drive, I was greeted by a sweep of intense color. In the spring, as the days grew longer and warmer, the garden erupted anew when the hundreds of bulbs I had planted, daffodils, hyacinths, tulips and wild irises all poked through the soil, dotting the hillside with blue and yellow. Neighbors, on their morning or afternoon strolls, would stop to admire the flowers. all over my house, vases bloomed with profuse bouquets.

The next winter, half the garden froze. Because drought-tolerant plants are heat-loving by nature and not resistant to cold, only my Peruvian lilies, along with a few salvias and sages, remained green after the freeze of 1990. A great deal of the rest, including my jacaranda, died. I began replanting immediately, the gazanias, the climbing jasmine, a second princess tree, the African daisies, as well as all my annuals. To my surprise, gradually the yarrow, the pansies, the penstamenons and more of the sages came back. By springtime, my hillside was once again ablaze.

Still, I continued planting, filling in even the smallest bare spot. One day at Home Depot, I discovered salpiglossis and miniature delphinium. I settled these in a pocket of large rocks to the side of my deck. Another time, I fell in love with a six-pack of pastel phlox. And another with orange and yellow mimulas, which I immediately envisioned creeping over my rock terraces. Finally, when I could find no more open spots to plant, I realized that my garden was complete. But I continued to spend hours outside. Now instead of planting, I gazed at my garden. From the stone bench I had sculpted into the hillside, I watched the humming birds extract nectar from the sages. From the deck, where I removed the railing so as not to obstruct my view, I painted, watercolors and pastels. In the morning, I'd bring my cup of tea outside and listen to the birds and smell the scent of hyacinth. In the late afternoon, I would watch the colors deepen, the air filled with jasmine.

The last time I saw this garden, I was leaving for Hawaii, and the wild ginger was in white and aromatic bloom. A week later, just as I was to set off for Mona Kea, my son Adam called to say that

Hiller Highlands was on fire. I didn't know for certain that the Gravatt Dr. house was gone until I called two days later from a pay phone at the crater rim (it had taken that long for my call to get through).

I returned to Berkeley to find out that both my gardens had burned. At Gravatt, I was greeted by my three tons of rocks and the skeleton of a single pine. The entire hillside was covered with damp, gray ash. (At Buckingham, I later saw, every single tree was burnt to the ground). Still, it didn't take me long to begin digging. At first it was to sift through the ash, where I discovered a new landscape of charred terra cotta pots, fused glass and twisted metal. Then, two weeks after the fire, I came upon a row of chives, two inches high, growing in what had been my herb terrace along the side of the house. Next I noticed the Peruvian lilies peeking through the duff. After this, I found myself returning to my hillside every afternoon. At first I went almost instinctively, not always aware of my destination when I set out. Gradually, I understood what motivated this daily pilgrimage. I needed to see the hillside over and over again, covered as it was with ash, to make the fire a reality. At the same time, I needed to nurture my garden in its efforts to reemerge. With each visit I would come upon new life. The blue sages multiplied. The Peruvian lilies, the Mexican marigold and the yarrow put out new shoots. Even the yellow snap dragons began budding. As my hillside slowly came alive again, I became protective of my plants, determined to eliminate the Scotch broom, the thistle, the anise which threatened to crowd them out. But if I was nurturing my garden, it was also nurturing me. Pulling weeds, turning over soil, I participated in the process of renewal. Slowly, I began to feel whole again.

Now it is spring, and at least once a week, I drive

from my temporary home in the Kensington hills, through Tilden Park and down Claremont Canyon to Gravatt Dr. There I pick, daisies and Peruvian lilies from my garden to make bouquets which I place on the window sill behind my kitchen sink. Sometimes on my way home, I swing by the Buckingham house, where all that remains are the steps leading down to a concrete pad still bearing the initials Adam wrote 16 years ago in the wet cement. Yesterday, I saw a bouquet of flowers the current owner had placed at the top of the drive.

First Cookies

Elisabeth Wickett

ooking is like making love: if you have to think about it, it isn't much fun. And so I put it off and off. Friends gave me an odd assortment of kitchen things: a meat loaf pan, a butter melter, frying pans, strainers, a pair of salt and pepper shakers. I put them away on neatly-papered shelves.

But last Thursday I had to make chocolate chip cookies. My students had analyzed the annual reports of Ben and Jerry's and Dreyer's and we were planning an ice cream party. What is ice cream without chocolate chip cookies?

I made a grocery list from memory. I didn't have the stand-by recipe on a bag of chocolate chips, nor did I have my own special recipe instantly available in short-term memory. I went to the store. I bought a five pound bag of flour, little boxes of white and brown sugar, a large resealable bag of chocolate chips—for encouragement—and six eggs. Then at the kitchen shop I bought a mixing spoon, measuring cups, ice cream scoop and grater for the orange peel. Butter—I had forgotten butter! And I had thought it was softening up in the grocery bag while

I shopped. I hopped into the car and went back for butter.

With my fire-impaired hands, I so diligently clutched the spoon to mash the hard butter that the olive wood snapped in half. Back to borrowed stainless steel. Baking soda? There was always a box of baking soda at the back of the stove in case of a little fire. But two boxes of baking soda, cardboard strong arm and quashing hammer, had not halted that raging firestorm eight months ago. Nor had the box of baking soda kept in the refrigerator. So, I hopped into the car once more and bought baking soda, an active ingredient, no longer a talisman.

With the dough mixed, I pre-heated my landlady's oven to 375 degrees. I set in the first two cookie sheets (borrowed). I burned the cookies. Ovens are just as idiosyncratic as people; I had forgotten to enquire about peculiarities. Of the rest, a couple of sheets were too dry, a couple came out almost right.

I'm fussy. Usually I'm fussy. With cookies, burnt, dry, and almost right, I had a triumph. Creating and baking only took eight months and four hours. First cookies *post incendium!*

from FIRENOTES

I notice how fire cleanses, transforms and brings change to the earth. My hillside took on the aura of a foreign cemetary, an archeological site. At first, nothing was familiar, and I felt empty of all feeling. Then, stories surfaced with images: terror in the eyes of a deer running past a person fleeing the flames, a woman in her driveway, turning to see her house melt into the foundation, two women in a swimming pool covering their heads with wet towels to breathe through as flames passed overhead. Finally, in the spring, the hillside was returned to the natural splendor of the season by the appearance of wave upon wave of wild flowers on the slopes still punctuated by the rows upon rows of charred trees. In fact, the spring display of wild flowers and flowering trees was the most beautiful in memory. I know that I will plant an orchard on our own fire-scarred field.

—Margaretta K. Mitchell

Ordinary Pleasures

Deborah Sasha Hinkel

April 9

ince the fire, I have rarely allowed myself ordinary pleasures. Indeed, I make every attempt to be alone, to be away from noise, the ordinary sounds of modern life. And although a month ago I bought a fabulous sound system, I have not been eager to try it or found myself psychically in need of it.

I do not watch the news, having substituted four newspapers for the noise of television.

I will do everything to avoid driving or even being with any people, except some fire survivors.

Why do I get solace in being alone at home? Why will I read only at home? Why is noise—mechanical sound—painful? Why can I not play the beloved Steinway or even listen to music? What keeps me at this address; a house not even mine? Why am I afraid—yes, that's the word, afraid to leave? To leave this house requires the fortitude of engaging in an expedition.

A lizard suns himself on my garden wall. All my senses are totally focused with this motionless reptile. When he darts off, I realize that I do not simply watch him. I am trying to hear. What am I listening for?

I think it is this:

The first crackle of fire. A shift of the wind. Those subtle sounds that are practically blaring in retrospect. If only I had been more watchful, more alert, I could have done something else.

If I had not been listening to Brahms, I might have heard the turning of leaves; the snap of twigs, the crumple of grass. I watched the wind blowing shingles off houses at nine-thirty that morning. I went out to water my garden. I was listening. But not carefully enough. If only I had demanded a fire patrol the night before. If only . . .

Terry Dupont and Bob Cox would still be alive.

Death is made of if onlys. The living must play the cards in hand now.

My housekeeper, Sandy, has arrived. She will give me three hours of Hemingway therapy, leaving a clean, well lighted space. It is a luxury I would never have felt I deserved before. I want to be at this house with my great Akita, Kushana. I want to hear each car roll by. Any helicopter overhead. Nothing to distract me.

Now the noise of clothes tumbles from the dryer.

I open a package; it is a CD Paul gave me for Christmas. I figure out the controls on the sound system, hook up speakers throughout the house, put on Bach's sixth Brandenburg Concerto. The sound swells in classic joy. It is the first time I have heard music in months.

Sandy can listen for the fire.

I begin to float on the careful counterpoints. Buoyed from gravity and guilt, my mind releases her grip and for a few moments my soul grabs for me that peace.

ELEGY

Sooner or later
we all compose a landscape for ourselves.

It might be
the lawn of the house where you were born

its birches
its Japanese maples, one rosebed, its pyracantha hedge.

Or the acreage
of an old farmer you settled near

his basement
shelved with currant jam and strawberry preserves

his yard
of wren houses, wild grapes and carrion plant

along the fence
the fields, the river banked with wild roses.

And later
wherever you see fields, or roses, you see

that same old man
stoop to pet his dog on a dusty road in Iowa.

You could have moved
ten times since then, own a house whose yard's bricked in.

Around you
Pacific rhododendrons bleed red.

Your neighbors
prune their Meyer lemon shrubs.

The old man is dead.
His acreage rented to the rural mailman as pasture.

His guides to Iowa
trees and wild flowers in his granddaughter's hands.

No matter.
Keys to prairie flora will not reveal the secrets

of coastal vegetation
where the exotic must become familiar for you to feel at home.

Where maples
give way to acacias, bloodroot to sourgrass.

Where it never snows.
No matter. The view from your windows is water.

You stare out
and the old man and his land fill the air, settling on you

lightly, like dust.

—Jane Staw

Looking Homeward

Nathaniel C. Comfort

Six months after the fire consumed our home, the fear and rage over what I had lost had ebbed, leaving, unexpectedly, surprise. I thought at first that I had lost my home, as in "I'm going home for the holidays." I had thought that my permanent dwelling was gone, the place I could run back to if worst came to worst, the place where family would always welcome me, no matter what foolish thing I had done.

But on reflection I realized I had relinquished that place long ago. I hadn't used Golden Gate Avenue as a "Permanent Address" since college. Throughout graduate school and now, out in the real world, my current and permanent addresses have been the same.

I t was not the home of my present that I lost, but the homes of my past and future. It was the comforting shell of my boyhood, and the fortifying possibility of bringing my own fledgling family back there someday.

As a young boy, I was no more conscious of my house than any other kid. But after my parents divorced, I necessarily took on more than the usual child's share of responsibility for the house, and with it came a caretaker's love. With my father gone, my mother and I each filled one leg of the proverbial family pants. My mother, sister and I painted the house's plaster walls and waxed its hardwood floors, my sister and I delighting in being allowed to put on dark socks and buff the planks by getting a running start in the kitchen and skidding through the hall, off the step into the living room. When I was about fourteen I spent two weeks sanding, staining and varnishing our thirty-odd kitchen cupboard doors and drawers. For years their warm, shiny surfaces gave me a surprising little jolt of pride when I returned home for holidays.

But my bailiwick was the garden. It was pretty bedraggled, but it and I became so intimate that I ignored its vast swaths of bald clay, and saw only the volunteer iris or fresh, pale clumps of grass. Occasionally I would receive a grant from my mother to replace the perennially moribund geraniums, or to buy some compost for the battle-scarred holly. I lovingly mowed the thin, oak-shaded lawn and shaped the evergreen ivy along the steps, I raked the constant fallout of live oak leaves from the sidewalk, path and driveway. Once a year, I climbed on top of the sloping cedar shake roof to clear the gutters of choking leaves; if I didn't, the playroom flooded in winter. These jobs began as chores, tasks that had to be done before I could go play basketball or go skateboarding. But in their repetition and the satisfying results they produced, they evolved into a sense of stewardship over our suburban homestead.

This feeling peaked during my last year in college. That was my mother's first year in Pittsburgh, so I moved back to the house and rented out rooms to friends and other students. I lived a fantasy that could have come true; as I ran the house that year, I thought of someday bringing my family back to this house, sleeping with my wife in my parents' room, tucking in my son and pulling down the shade of the window I had stared through as a child. While my peers used their spending money for record albums and beer, I bought bathtub stoppers, fuses, and mulch.

One of my proudest purchases that year evolved into a half-joking symbol of our far-flung family's connection with the house. I bought a peach sapling.

I ferried this wispy charge home from Pay Less, selected a bright spot in the back yard, dug a good hole in the rocky adobe hillside, filled it with mulch and woodchips, and tucked in the little tree. Someday, I could send my son out to the back yard in bare feet on a late summer Sunday morning to pick enormous, juicy peaches with which to smother our waffles.

Well, either that tree was a dud, or I wasn't half the gardener I fancied myself to be, but after five years the tree had borne a grand total of three peaches. Three times my mother, who made the most frequent visits back to the house, called gleefully to tell me there was a peach on that rangy twig. With a little help from our emerald-thumbed neighbor, whose hose just reached the tree, that tenacious little peach squeezed out a handful of leaves and a miraculous fruit every year or so. Sappy as it may seem, the unspoken assumption among us was that the tree's flimsy roots were binding us to the house, promising us its sweet fruit. After the fire, my mother arrived in the Bay Area two days before me and immediately went up to see the ruins. That night she called, laughing and crying at the same time, to tell me that, incredibly, the little tree had survived. That was the strange thing about the fire; often trees remained green, and even flowers bloomed, while twenty feet away it had been hot enough to melt wrought iron.

When I graduated from college, I left the house and the Bay Area for graduate school. My sister, still in college, took her turn as its caretaker. After the fire, I saw that she loved the house perhaps more than any of us, but when I returned that year to visit, the garden was desiccated and the house seemed run-down. When I pleaded with her to take pride in the house, to treat it as family, she responded with great insight: "I want to have my own house to take care of." She didn't share my dream of raising children in the house. Her vision of adulthood held a fresh new house, unhaunted by memories, in which she would create a home from scratch.

We are rebuilding now, and the three of us hold cross-country conference calls to discuss architectural sketches for the new structure. We tend to argue about how many elements of the old house should be incorporated into the new. My mother and I favor including some of the lovely detailing that made our house, in my unbiased opinion, the prettiest on the block. My sister, though, says she would feel surrounded by ghosts if they were included. She has training in architectural design, and is probably right about modernizing the house. Rationally, I know we will never have the old house again. Duplicating it would only make me search for long-vaporized plaster cracks and listen for muted midnight floorboard creaks. But part of me can't avoid feeling that replicas of our old, beamed cathedral ceiling, tiled stairway landings, arched alcoves, or even the crummy linoleum kitchen floor, would restore a shred of the sense of lineage that comes from knowing I could, if I wanted, finally sit at the head of my childhood dining room table.

Zones

Constance R. Rowell

efore the firestorm, I had never picked up a camera. A week afterward, I would not be without one. I did not know at the time what motivated me, nor did I stop to ask. I was in too much of a hurry to capture on film what had captured my eyes, as if this new and solemnly transformed universe might disappear at the wave of a wand, leaving no trace of its tormented beauty, of the phantasmagoric landscapes that lay like a mirror upon my mind.

Ill at ease outside the fire zones, I spent my days exploring them, coming back to certain sites again and again. Reckless, I trespassed, ignoring warnings, venturing inside buildings that were declared unstable, sneaking around like a thief, making a pest of myself.

In the open, walking along the streets, I talked to anyone who was willing: to work crews, to police officers, to sightseers (looky-loos as the cops called them), to insurance agents, to burned-out homeowners. We traded stories. In the case of the latter, we traded tears. We made comparisons: World War II, Berlin, Hiroshima. No one, however, seemed to see what I was seeing: Rome, Florence, Jerusalem.

One day, sneaking past a police barricade on Alvarado Road, I walked up a flight of steps and found myself in the midst of a magnificent, fire-gutted estate; some of the inner courtyard, with its pool and fountain, was still intact. In one corner, luminous against the charred earth and blackened trees, a fertility goddess stood in Botticellian pose, sheaves of wheat held loosely in one arm, head tilted demurely to one side.

Afternoon shadows began to make their way across the white, ruined walls. The day felt quiet and warm. I lingered for a while by the pool. Two three-story chimneys from the adjoining property, their sides glazed with light, shone in the dark water like slants of silver. Drawn to their glinting images, I took a black and white picture of their reflection in the pool.

A week later, print in hand, I studied it, mesmerized by the bits of flotsam that fluttered on the pool's surface, by the chimneys themselves, ghostlike and ethereal. The whole effect, though beautiful, was ominous. That night, I had my first fire-related dream.

I was a teenage girl, playing with my best friend, Happy Jordan, who lived in a big house on Alvarado. We decided to go exploring and entered a deep thicket. Coming out on the other side, we found ourselves facing a dark, brooding pond. We saw some neighborhood boys on the opposite end, throwing stones, but we didn't pay them any heed.

Happy, without a word, waded into the pond. I watched as she began to sink, then disappear from sight. My mind, locking into slow motion, seemed to quell the world around me. Leaves sat on the water's surface. The day was without sound. Struggling to break from my trance-like state, I called to the boys—*help*—*help*—*help*—my voice an echo that floated weakly on the air. They continued with their games. Unless I did something immediately, my friend would drown.

As I entered the pond, the water's chill against my bare legs made me gasp. In that one second I knew, as if I had become my real age, that if I plunged in after her, I had neither the strength nor the breath to return alive.

I wondered how I would tell her parents that she was dead, that I had been too afraid of losing my life to save hers.

A Walter Mitty fantasy of mine, as an adult, was that I would brave a fire if it ever came to where we lived, at the dead end of Balsam Way. In this heroic version of myself, I remain behind, while all others flee, and fight off the flames with my garden hose, saving the house, saving the neighborhood. Never in my imagination would I follow police orders to evacuate. Yet that is what I did on October 20. I heard, I saw, I fled.

Later that afternoon, I did sneak back on foot, closing the front door that we had left open (for the sake of one of our cats who had hidden to escape being put out), turning off the hoses that we had left running, dashing back down the hillside to where I had parked my car on Pinehaven Road. Yet I could have stayed. No one was there to order me not to. Helicopters and fire crews were too busy fighting the flames on nearby Broadway Terrace. So I could have kept a neighborhood watch. I could have drenched our end of the street. I could have . . . but I did not. In fantasy a heroine; in reality a coward.

I continued to travel the fire zones with my camera, spurred by the knowledge that these areas would soon be cleared. I was in a state about it, having fallen in love with many of the sites, having visualized them to an extraordinary degree, endowing them with an agony as ancient as the Crucifixion, a ruined grandeur that equaled Rome, a cast of colors as richly hued as those imbedded in the sun-baked walls along the Arno.

Inwardly, however, I was falling apart. I broke into tears for no apparent reason. I lashed out at my husband. I readily became hysterical. The smallest responsibility left me on the verge of a breakdown. I photographed by day. At night, I was undone.

Closeting myself in the bathroom one evening, not knowing how else to hide my disintegration, I stared at myself in the mirror and blurted out: like a god gone wild, like the dead reliving their dreams. Suddenly I was alert. But what dream? For I now understood that what had prompted this outburst was a dream I once had—long ago—a dream about having been dead—or almost.

In touch with old memories, I watched them unfold. I saw our house at 255 The Uplands in Berkeley. I heard air raid sirens. I saw my mother running to let down the thick, black curtains, then hurry us off to our "stations" in the basement, my brother and I in our cots, close together, my sister

194

and mother farther off.

Then planes came. Bombs dropped. Minutes later, I found myself in a world of ruins, our stucco house reduced to rubble. No longer "us," however, no longer "our." My brother, my mother and my sister had all been killed. The knowledge of how alone I was grew terrifying. Clearly, I could not bear it, for trotting toward me, as if out of the upper right corner of a photograph, came Chamois, our Australian shepherd. Some part of my dream-mind had sent him on a rescue mission.

Sometimes the understanding of what drives us dulls our pursuit. This was not my experience. Or perhaps my obsession matured into a more balanced love affair. I am rarely without a camera. The world I tried to interpret in my photographs remains. Yet it is to my Florentine tower on Cochrane Avenue (so named because its silhouette brought back to mind the Piazza della Signoria in Florence) that I owe my first allegiance. Against a rich November blue it shone one day, the metal flashing, bathed by the late afternoon sun, turning to a della Robbia yellow. Its sides, stripped of sheathing as flesh flayed of skin, yielded to brick and bright mortar. Every element in its composition stood out against the pure, dark backdrop of sky. Since then, nothing has diminished what I saw on that color-saturated afternoon.

from FIRESTORM

Spring now.

On the chimney pot   a small bird
in her mind   a nest

Your place of devastation
obsesses me.   There is nothing new there—
a white gazebo behind the absence
of a splendid house,   ivy
needing to be cut back,
the singular green
that mats over graves.

                       —*Nona Nimnicht*

CONTRIBUTORS

Patricia Adler is a collaborator, writer, editor and cartoonist. The fire was headed for her house, but the winds died and spared her neighborhood.

Jaime Benavides is a staff photographer for Kaiser Permanente. He was returning from a golf tournament when he saw the smoke.

Gregory M. Blais is a novelist. His wife, Betty, created a garden at their Broadway Terrace home that was featured in Sunset Magazine's "Drought" issue. The garden, along with their house, was destroyed.

Sean Bonetti is a firefighter for the San Mateo Fire Department. For six weeks after the fire he coughed up particles of soot.

Marion Abbott Bundy is in the Creative Writing program at Mills College. She lives with her family in the Elmwood district of Berkeley.

Nathaniel C. Comfort grew up on Golden Gate Avenue near the "Big Tree." He is now a science writer at Cold Spring Harbor Laboratory, Long Island.

Jack Compere and his wife, **Marvel Vigil**, live in the Montclair district. They both work for Pacific Bell.

Ellen Cooney lives in San Francisco, and has published four volumes of poetry. The day of the fire she was visiting friends in Oakland.

Bonnie Cox has been an Oakland firefighter for six years.

Laurie Dornbrand is a geriatrician. The morning of the fire she took her son to Children's Fairyland. They never got home again.

Deirdre English is a Bay Area writer and former editor of Mother Jones Magazine. She lost everything in the fire.

Sharon Fain lives in Piedmont and is a counselor, child development instructor and poet. Her poems have appeared in numerous small magazines.

June Felter is a Bay Area painter. The fire destroyed all but a handful of early figure paintings that survive in private collections, plus several dozen watercolors that she grabbed as she fled.

Linda Fletcher lives on Elmwood Avenue with her husband and three daughters. She writes, watercolors and does community work.

Steven Frus is a photographer who has lived in Berkeley since 1979. He used to bicycle through the fire zone every day on his way to work.

Taylor Graham is a volunteer search-and-rescue dog handler. His writing has been widely published.

Jesse Grant is a student at Oakland Technical High School. His family lost everything in the fire, but Jesse saved his camera.

Louise Gund is a photographer who lives in Oakland. She saved two cameras from the fire, but lost her entire portfolio.

Deborah Sasha Hinkel is a lawyer practicing in Berkeley. She lost everything in the fire, except her dog, Kushana.

Mary Hutton is a retired professor and a writer who has published in numerous literary magazines. She has lived in the East Bay almost all her life.

Susan Ito is an Oakland writer and an MFA candidate at Mills College. Her family's home was several blocks from the fire zone.

David Kessler is a librarian at the Bancroft Library. He lost everything in the fire.

Karen Klaber is the publisher of The Monthly. She lost everything in the fire.

Lynne Knight is a Berkeley writer who teaches English at community colleges. She has published in numerous literary magazines.

Jeremy Larner is an Oscar-winning screenwriter and poet. He lost everything in the fire.

Ann Leyhe lives in the Elmwood district of Berkeley with her husband and three children. She is a contributing editor for Horticulture magazine.

Reverend Samuel J. Lindamood has been pastor of the Piedmont Community Church for thirty years. He lived in Hiller Highlands, and lost everything in the fire.

Nora Mielke has an M.F.A. from California College of Arts & Crafts. She has lived in Berkeley for 35 years.

Margaretta K. Mitchell is a freelance photographer and writer. **Fred Mitchell** specializes in residential real estate. They raised three daughters in their Claremont district house.

Joan McMillan is a mother, quilter and poet who lives in the Santa Cruz mountains. She has published poetry in numerous magazines and anthologies.

Trent Nelson is a photojournalist for the Contra Costa Sun. He has lived in the East Bay all his life.

Nona Nimnicht is a poet living in Oakland. She has published in a number of small magazines.

Sharon Olson is a poet and librarian in Palo Alto. She has close ties to Berkeley through friends and relatives.

Luis Alfonso Paez-Cano is an historian who takes photographs to document history. He moved to this country from Colombia seven years ago.

Nancy A. Pietrafesa lives in Berkeley with her husband and their three sons. The fire stopped at the backyard of the house next door to theirs.

John Pollock is a professor of English at San Jose State University. He is the brother-in-law of Nancy Pollock.

Nancy Pollock is a Berkeley artist and realtor. She lost everything in the fire.

Frances Rowe came to Berkeley as a university student, fell in love with the area, and has lived in the Elmwood district for 28 years. She writes nearly every day of her life.

Constance R. Rowell *is a scholar, writer, and now a photographer. She grew up in Berkeley, and lives in the Oakland hills, on an unscathed block in the middle of the fire zone.*

Lori Russell *is a writer and part-time home health care nurse. She has lived in the East Bay for over 25 years.*

Sonia Saxon *has published poetry in a number of literary magazines. She lives in Redwood Garden apartments, and watched the fire from the Berkeley Pier.*

Terry Shames *is a writer. She lives with her son and husband in Berkeley, near the Claremont Hotel.*

Tobie Helene Shapiro *is a writer, cellist, composer and artist. She lost everything in the fire.*

Deborah Simpson *is an accountant who enjoys writing. She lost everything in the fire.*

Emily Jurs Sparks *lives in Crocker Highlands. Her parents still live in the Montclair house where Emily and her sisters grew up.*

Susan G. Spoelma *is a nurse who works in administration. She has a son, Alex, who is 20 years old.*

Jane Staw *lived in the Elmwood district for 12 years. She is a writer and teacher of writing.*

Howard Stein *is a film editor.* **Dorothy Stein** *is a caterer. They still live off Upper Broadway Terrace in their home spared by "heroic efforts" to contain the fire.*

Merle Stiles *is a photographer who is working toward her M.F.A. at San Francisco State University. She lives in Oakland with her husband, Charles.*

Peter Strauss *is a psychotherapist who practices in Danville and Castro Valley. His home survived the fire.*

Judith Stronach *is a journalist who lives in Berkeley. She has just published* Visible and Vulnerable: The People We See on Berkeley's Streets, *for sale by the homeless.*

Rick Talcott *is a lawyer who practices in Oakland. He lost everything in the fire.*

Leslie Tilley *is a writer and editor. She was born and grew up in the hills.*

Bernadette Vaughan *is an actor and writer. She heard from her son at 7:00 the evening of the fire—he had been sailing on the Bay, from where he saw his home, the Parkwoods Apartments, burn to the ground.*

Stan Washburn *is a trustee and teacher at The College Preparatory School. He is a painter by profession.*

Norman Weinstein *is a poet and critic who lives in Boise, Idaho. He used to live in the Bay Area, where he visited shortly after the fire.*

Mark Wholey *is a sculptor. He was flying back to Oakland in a Cessna 72 the evening of the fire. The horizon was completely red.*

Elisabeth Wickett *is a writer and teacher. She lived at the Parkwoods Apartments for seven years.*

Anne Ziebur, *whose poems have appeared in local anthologies, is a volunteer for California Poets in the Schools. In the fire, she lost the family home where she had lived since 1963.*

Cover Photo: Richard Blair, *is a commercial and fine arts photographer who has had a studio in Berkeley for almost 20 years. On the day of the fire, with only five shots remaining, he took the cover photo.*